PLANETARY HEREDITY

by Michel Gauquelin

Foreword by Professor Giorgio Piccardi
University of Florence, Italy

"We must keep our freedom of mind and must believe that, in nature, what is absurd according to our theories is not always impossible."

Claude Bernard
An Introduction to the Study of Experimental Medicine

PLANETARY HEREDITY

Michel Gauquelin

Cover Design by Maria Kay Simms, San Diego, CA

Printed in the United States of America

Published by ACS Publications, Inc.
P.O. Box 16430
San Diego, CA 92116-0430

Acknowledgments

I am grateful to Françoise Gauquelin for her invaluable help with the gathering of birth data in the first series of planetary heredity experiments.

I am also indebted to Dr. Geoffrey Dean (*Recent Advances*, Subiaco, Western Australia) for helpful comments and for corrections of my faltering English.

Contents

Illustrations and Tables

Figures

Tables

Foreword

(From the original Foreword by Prof Giorgio Piccardi written in 1965) A few years ago it would not have been possible for a serious scientist to write a foreword to this book. It would have been dangerous for his reputation; the subject is much too close to the occult. Today it is different. At last we are gaining some understanding, from a general point of view, of what may lie behind Michel Gauquelin's discoveries.

Fifteen years ago space was considered to be void, cold and inert. Between the celestial bodies there was no force but gravitation. Light crossed space without warming it. How could this cosmos, full of nothing, influence living beings? All was considered immutable. Solar radiation, magnetism and gravitation were alleged to be constant, fixed, unchanging. On Earth, other than the rhythm of the tides, nothing that happened was attributed to space. Planets and other celestial bodies were said to be too tiny or too distant to affect us, even in a very small way. So to talk about planetary heredity would have been senseless, superfluous, dangerous and stupid.

Fortunately, my life has been long enough to witness changes. Scientists today have filled space with matter and energy so that it now seems completely different. But because these changes are only recent, scientists still hesitate to consider how the cosmos might influence us on Earth. In fact a new field of research has been born: the field of fluctuating phenomena.

Fluctuating phenomena is the name we have given in my Institute of Physical Chemistry to chemical reactions that change from one day to the next, even though they are carried out under apparently constant laboratory conditions. Until now, scientists have tended to ignore these changes. but to us they are extremely interesting.

The study of fluctuating phenomena needs special methodology. Fifteen years ago we devised an approach based on repeating a standardized chemical test over a long period of time. Our tests have been run daily since 1951, and during the International Geophysical Year were used throughout the world in places like Africa, Japan and Antarctica. As a result, we have discovered unexpected effects from space; for example, our chemical reactions are affected by solar flares, sunspots, and the position of the Earth in its spiral path through the galaxy. We also found that biological reactions correlate with chemical reactions, showing that both are influenced by the same external causes. Thus we have added something to the science of life.

The study of fluctuating phenomena suggests that the universe has an effect on any given point in space. So it is not surprising that Gauquelin's careful study of planetary effects, using a strict scientific methodology, may have discovered a new link between humanity and the universe. One may ask: why is there a planetary heredity? My answer is: a lack of it might be the more unusual.

Gauquelin is to be praised for investigating planetary heredity despite serious professional opposition from the scientific establishment. He demonstrates that planetary effects on our life may be no different in principle from any other cosmic effect. Thus his work goes beyond the simple laws of statistics and the mechanisms of heredity; it adds to the great ideas that have emerged from studies of the relations between mankind and the universe. We should be happy that we may have a new link in the chain of knowledge that connects our life more closely to the whole.

Prof Giorgio Piccardi
Director, Institute of Physical Chemistry
University of Florence, Italy
June 20, 1965

Publisher's Foreword

The sun and moon are lower-case when used astronomically and capitalized when used astrologically. Exceptions are Figures and Tables.

ACS Publications
May 1988

Preface to the English Edition

My experiments on planetary heredity are often inaccurately reported, probably because the important early work was not published in English. Thus the hypothesis was outlined in 1960 in French,[1] the methodology and first results appeared in 1961 in German,[2] and in 1962 in French.[3] In 1966, *L'Hérédité Planétaire* (*Planetary Heredity*), my most comprehensive book on the subject based on 25,000 birth data, was published in Paris. Twenty-two years later I am happy to present, at last, an English translation of this book.

The original French manuscript was written in 1962, then revised in 1964, but due to printing delays it did not appear until January 1966. For this American edition a number of changes have been made. The text has been condensed where necessary to improve readability and to exclude speculations that, during two decades, have become obsolete. A chapter on biological clocks and cosmic influences has been omitted; this information can be found in more detail in *The Cosmic Clocks*, first published in 1967 and republished in 1982 by ACS Publications. More importantly, a chapter on astrological heredity has been added together with a postscript summarizing the developments in planetary heredity since 1966.

The original French edition contained a long and enthusiastic foreword by Prof Giorgio Piccardi, famous for his investigations of extraterrestrial effects and for his discovery of the Piccardi effect. Because of the advances in scientific knowledge since then, and because the book has been updated, Prof Piccardi would have wanted to update his foreword also, but sadly he is no longer alive to do so. So his foreword has been condensed to exclude dated material.

The text is aimed at the general reader and is therefore free of heavy technicalities. Comprehensive details for the technical reader are given in appendices. It is my hope that this new edition will be both a guide for the curious reader and a scientific document for the scientist who wishes to examine the complete evidence for planetary heredity. An English translation of the most important parts of my earliest non-heredity work[1,5] is also available (*Written in the Stars — The Best of Michel Gauquelin*, Aquarian Press, UK 1988), so English-speaking readers now have access to all my first discoveries.

I consider **planetary heredity**, if real, to be a potential Rosetta Stone for all my findings. It represents the best opportunity to integrate into modern science the "neo-astrological" correlations I discovered between certain planets and famous people. Planetary heredity is also important because, if real, it shows that planetary correlations exist for ordinary people as well as for famous people. There is, in principle, almost no limit to gathering data on ordinary people, contrary to what happens with famous people, a group strictly limited by definition. This abundance allows us to disentangle complicating variables such as induction and geophysical factors. Consequently, although the correlations due to the planetary heredity are somewhat weaker and less spectacular than the correlations found with famous people, they may in the long run be far more important.

Paris, June 1987
Michel Gauquelin

References to Preface

1. M. Gauquelin (1960), *Les Hommes & les Astres*, Denoel, Paris: 232.
2. M. Gauquelin (1961), Die Planetare Hereditàt, *Zeitschrift für Parapsychologie und Grenzgebiete der Psychologie*,5,2/3:168-193.
3. M. Gauquelin (1962), Existe-t-il une hérédité planétaire? *Planète* , 6: 77-83.
4. M. Gauquelin (1962), L'hérédité astrale, *Les Cahiers Astrologiques*, 98: 135-143.
5. M. Gauquelin (1955), *L'Influence des Astres, étude critique et expérimentale*, Dauphin, Paris.

CHAPTER ONE

THE PUZZLE OF CHILDBIRTH

To understand the observations presented in this book concerning cosmic factors and human birth, we should first look at present medical knowledge concerning delivery and birth. What are the mechanisms of delivery? What role does the fetus play? What kind of influence could cosmic and environmental factors have? For me, these are very important questions. But they are generally received with almost complete indifference by those who should be most concerned, namely astrologers. They place great importance on knowing the exact time of birth in order to erect a comprehensive birth chart. But oddly enough they show very little interest in the delivery itself, even though this underlies the timing of birth. What the fetus does before birth is of no concern to astrologers, and they regard it as an object without cosmic sensitivity. It is only outside the mother's womb that it becomes a person on whom the stars will have conferred a soul, a character and a destiny at the moment of the first cry. I do not share this lack of astrologers' interest. To the contrary, I think that a better knowledge of the physiology of childbirth may provide helpful clues toward an explanation of the planetary effects.

Delivery Description

The human delivery is a series of physiological phenomena that take place at the end of the ninth month of pregnancy. It starts when the first pains occur due to the contraction of the uterus; in medical terms, this is the "onset of labor."

Classically, the delivery is divided into three stages: the dilation of the uterus, the expulsion of the fetus, and the delivery of the placenta.

The first two stages are called "the labor"; the last stage is called "the afterbirth." The first stage, the dilation of the cervix, is the longest, and usually takes several hours or more (see below). The second stage begins when the head becomes visible, ends when the baby is born, and usually takes less than twenty minutes. In medical terms, birth occurs when the umbilical cord connecting mother and baby is cut. Then, generally some twenty minutes later, the placenta is discharged from the uterus of the mother.

The duration of labor varies so much from one case to another that the average (about eight hours) is indicative only. Numerous factors can modify the duration or difficulty of labor, for example, race, lifestyle, individual differences due to the mother (age, physical constitution, number of previous births, etc.) and individual differences due to the fetus (weight, position in the womb, etc.).

Medical progress in childbirth has shortened the average duration of labor. In 1900 a delivery lasted about fifteen to seventeen hours for primiparous (first baby) mothers and about six to eight hours for multiparous (subsequent babies) mothers. Today, in Western countries, the average is about seven hours for primiparous mothers and four hours for multiparous mothers, but still with wide variations.

Causes and Mechanisms of Delivery

Delivery is under control of the uterine contractions. In his book, published in 1951, the French doctor H. Vignes indicated that the average number of contractions is one hundred and twenty for primiparous mothers and seventy for multiparous mothers. The closer the end of delivery, the more frequent and painful the contractions.

Mechanisms are well described. But causes remain largely unknown, as the renowned obstetrician Prof R. Merger frankly pointed out in his *Précis d'Obstrétrique* . Physicians have had problems in identifying the physiological factors that trigger labor and set the rhythm and intensity of uterine contractions. For example, the specialist J. D. Ratcliff says, "The onset of labor is one of the great unknowns of medicine." Until recently, and despite numerous experiments and theories, nobody had succeeded in definitely answering questions like: is the cause of labor purely mechanical or is it due to the nervous system or some hormonal factors? Dr F. Dessagne, in his medical thesis, observes: "parturition is the upsetting of a hormonal equilibrium which was established since the pregnancy." Here it is worth stressing that hormones can be active at infinitesimal doses for example, as J. D. Ratcliff says, "One part adrenalin in 400 million parts of blood is enough to cause specific reaction in a human being''; and it is well known in obstetrics that it may take only the tiniest thing to trigger the labor off. This explains why investigation is so difficult, and why physicians favor some sort of hormonal implication in delivery processes.

The role of the fetus is still unclear. Not very long ago, the majority of physicians thought that the fetus is passive during delivery, and that only the mother's body is active. But, progressively, the importance of the fetus is being acknowledged. Now they are not far from accepting Hippocrates' assertion of some 2,500 years ago that: ''when the time is ripe, the child moves, breaks the membranes holding it, and leaves its mother's womb.'' Of course, the specialists are now thinking in hormonal terms, not in naive mechanical ones. In other words, when the time has arrived, the fetus releases hormones via the umbilical cord into the mother's bloodstream. These hormones reach the uterus and cause it to contract. This continues as long as the fetus is connected to the mother by the umbilical cord; when the cord is cut, contractions cease. This view is supported by the discovery that traces of fetal blood can be found in the mother's blood during and after the delivery.

My own findings suggest that the cosmos may affect the fetus in some way during the delivery. The discovery of an active role for the fetus in timing its own birth is therefore a strong argument for a possible scientific explanation of my findings, or at least a favorable inducement for seeking such an explanation.

Extraterrestrial Factors and Diurnal Rhythms

According to some authors, the frequency and quality of deliveries are affected by geophysical and astrophysical factors. Thus, in 1952 the German physicist R. Reiter found an increase of births on days when very low frequency radio waves were present in the atmosphere, or when geomagnetic activity was quickly changing, both phenomena being caused by solar flares. Two years earlier another author, W. Cyran claimed that onsets of labor are more frequent during geomagnetically disturbed days. In his 1956 medical thesis, J. Meyer noted that the minimum of births tends to occur on the day before outbursts of geomagnetic activity, with the opposite tendency two days afterward. On the other hand, authors who tried to link the monthly lunar cycle with the frequency of births did not find consistent results; some of them, like H. Hosemann in 1950, R. Reiter in 1952, and J. Meyer in 1956, found no lunar effect at all. By contrast a very clear diurnal rhythm of births has been demonstrated by a large number of statistical experiments. Early studies like those of V. Goehlert in 1888, and more recent ones like those of M. F. Gauquelin in 1959, all show very strong evidence for such a 24-hour rhythm. In humans, there is a broad maximum of births around 3:00 - 4:00 AM, and a broad minimum around 1:00 - 2:00 PM (Figure 1 bottom). There is also an even stronger 24-hour rhythm for the onset of labor; for example, H. Kirschhoff in 1935, E. Charles in 1953, and J. Malek in 1962 all found a maximum around midnight and a minimum around noon; and I found the same for 3,051 labors in Créteil (a Paris suburb) Hospital for deliveries occurring between 1938 and 1945 (Figure 1 top).[1] The amplitude of the onset rhythm in Figure 1 is roughly ±20%, almost twice that of the birth rhythm. Malek also found that if the onset of labor occurs when it is most frequent, that is around midnight, the delivery is easier and shorter than if the onset occurs when it is least frequent, that is around noon. Here we have a good example of the strong 24-hour "time giver" that affects our biological and psychological activities.

Modern Deliveries

Today we are observing a complete distortion of the basic rhythm of births. It began at the beginning of this century, but has been readily visible only since 1950 or so, and is getting more and more obvious. The main cause of the distortion is due to the artificial induction of labor. Induction was, for instance, described as early as 1936 by R. Vaissiere in his thesis and is now a very widespread practice in which numerous medications are used to induce labor or to stimulate its progress. Of course, around 1950 and probably still now, the use of induction depended on the doctor and on the maternity ward. That is why, as M. F. Gauquelin noted in her 1959 study, the daily pattern of birth differs not only according to decade but also according to location. In some cases, increasingly frequent today, the natural rhythm of birth completely disappears, and a "medical" rhythm takes its place showing a maximum at the middle of the day. In certain cases, births occur almost exclusively during business hours (Figure 2).

Whatever we may think about the transformation of our ancestral way of being born and regardless of the benefits that it may bring in necessary cases, induction upsets the normal biological processes of delivery. For me, it also upsets the chance of explaining strange planetary effects at birth, so I must regret this tendency toward "delivery

1. M. Gauquelin (1967), "Note sur le rythme journalier du début des douleurs de l'accouchement," *Gyn. Obst.* (Paris), 66, 2, pp. 229-236.

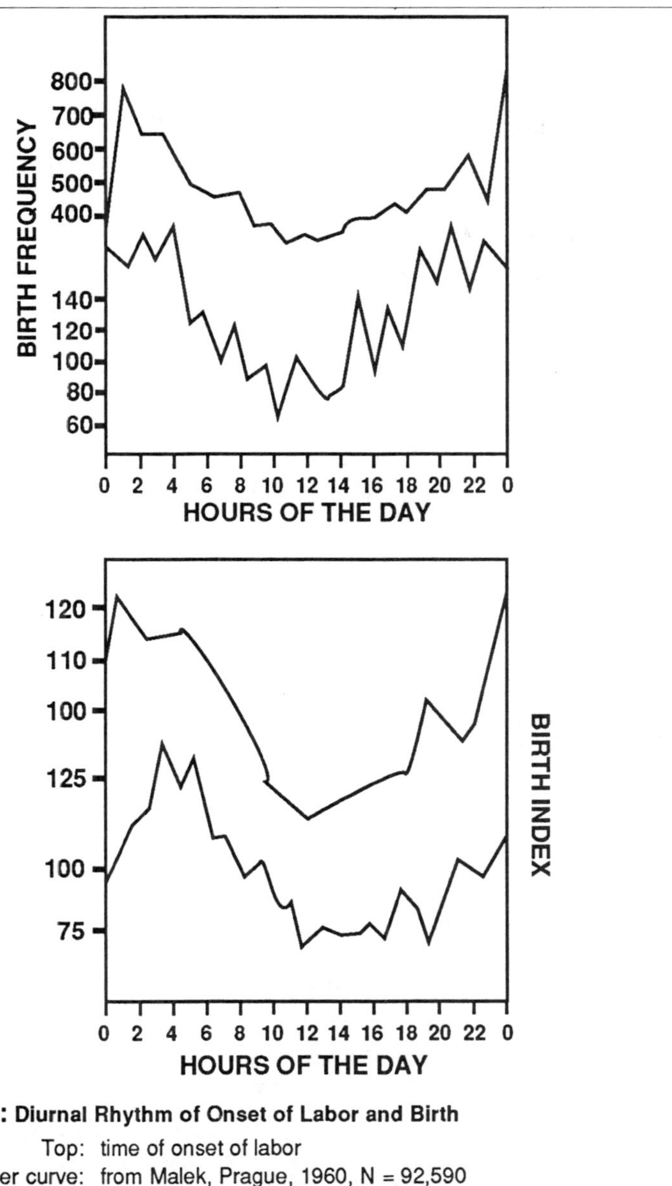

Figure 1: Diurnal Rhythm of Onset of Labor and Birth

Top: time of onset of labor
Upper curve: from Malek, Prague, 1960, N = 92,590
Lower curve: from Gauquelin, Creteil, 1937-1945, N = 3,051

Bottom: time of birth
Upper curve: from Goehlert, Zurich, 1876-1888, N = 86,850
Lower curve: from Gauquelin, Paris, 1928-1942, N = 3,124
Birth frequencies have been adjusted to mean = 100

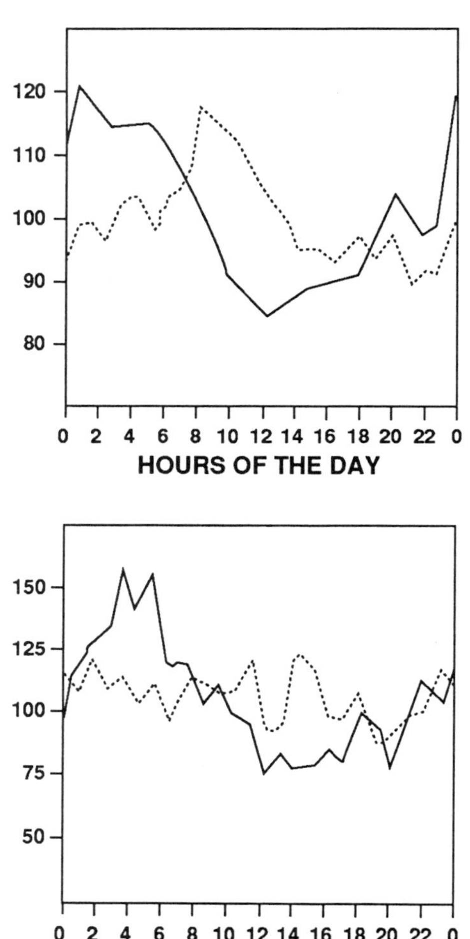

Figure 2: Transformation over the Years of the Natural Daily Rhythm of Births into a Medical Rhythm

Top: Birth index for the same country Switzerland. Solid line, Goehlert's observations (1876-1888). Dotted line, Fasler's observations (1929-1938).

Bottom: Birth index for the same maternity ward, Saint Antoine hospital, Paris. Sold line, births from 1928-1942. Dotted line, births from 1943-1957 (from F. Gauquelin, 1959 & 1960).

by appointment.'' We therefore have to be quick to observe and, if possible, explain the effect of the cosmos on normal births, before it is too late.[2]

2. *Note of 1988*: Of course, the matter of this chapter was written in 1966. Readers may find updated information about the fetus role in timing birth and on modern deliveries in Michel Gauquelin, *Cosmic Influences on Human Behavior*, Third Edition, Aurora Press, New York (1985), Chapters 15 and 16, pp. 187-203; and in Michel Gauquelin, *The Truth about Astrology*, Blackwell, Oxford, 1983 (published in the U.S. as *Birthtimes: A Scientific Investigation of the Secrets of Astrology*, Hill & Wang, New York). Chapter 9, pp. 160-179, gives the latest implications and discoveries on the subject.

CHAPTER TWO

BIRTH TIMES OF FAMOUS PEOPLE

In his famous book *An Introduction to the Study of Experimental Medicine*, the French biologist Claude Bernard says: "Experimental ideas are often born by chance, with the help of some casual observations. Nothing is more common; and this is really the simplest way of beginning a piece of scientific work. We take a walk, so to speak, in the realm of science, and we pursue what happens to present itself to our eyes. Bacon compares scientific investigation with hunting, where the observations that present themselves are the game. Keeping the same simile, we may add that, if the game presents itself when we are looking for it, it may also present itself when we are not looking for it, or when we are looking for game of another kind."[1]

This remark by Claude Bernard is in accordance with my own scientific adventure. At the beginning, the game I was hunting was not exactly the same as I was catching in my nets. I was checking the claims of traditional astrology, and partly by chance, I discovered a very strange correlation between planetary positions and the birth times of famous people. In this chapter I would like to briefly describe my first observations as published in my first French books *L'Influence des Astres* (1955) and *Les Hommes & les Astres* (1960).

The Sky and the Diurnal Movement

Let us imagine that we are in flat countryside under a clear night sky. Looking up, we see the constellations, which are made up of stars and

1. Claude Bernard, *An Introduction to the Study of Experimental Medicine*, Part III, Chapter 1, Dover, New York, 1957, page 151.

bear names that are familiar to us such as the Big Dipper, the Little Dipper, the Dragon, Hercules and Orion. We can also see the moon and certain planets. The planets generally appear brighter than most of the stars, even though they are much smaller, because they are part of the solar system and are therefore much closer.

It is just midnight, and the air is cool and fresh. The moon is rising on the eastern horizon, Mars is directly above us, and Jupiter is about to set in the haze on the western horizon. The other planets are below the horizon and are therefore not visible. Four hours later the configurations have changed. The stars, moon and planets have all moved in the same direction by the same apparent amount. For instance, the Big Dipper has moved toward the west, the moon is now high in the sky and Mars is now close to the horizon. Jupiter has disappeared and is continuing its journey below the Earth. Venus has risen over the horizon, heralding the sunrise. But these are only apparent movements, due to the Earth rotating on its axis every twenty-four hours in a uniform motion. We all know that this rotation is responsible for the alternation of day and night. Every day the same phenomenon occurs among all the stars. Thus the sun, moon, planets and constellations rise, reach their highest point in the sky (upper culmination), set, reach their lowest point below the horizon (lower culmination), and then rise again the next day, all due to the Earth's rotation. But each follows its own schedule in time and space.

12 Astrological Houses vs Gauquelin 36 Sectors

For at least two thousand years astrologers have made use of diurnal movement. Their astrological ''angles'' (Ascendant, Midheaven, Descendant, and Imum Coeli) correspond to the astronomical rise, upper culmination, set and lower culmination points, and are alleged to have a great significance for the interpretation. Astronomers have for a long time been able to calculate, in advance, the exact time at which the sun, moon and planets will rise, culminate and set. Their calculations are published every year in astronomical ephemerides for each country. Suppose we wish to determine the movement of Mars observed in Paris on May 24, 1956. From a French astronomical ephemeris for that year we find that Mars rose at 0:44 AM, reached its culmination at 5:33 AM, and set at 10:22 AM. In the same way it is possible to determine the diurnal position of the sun, moon and planets very precisely for any location and for any given date.

Now, suppose that on May 24, 1956, a baby was born in Paris. If he was born at 1:00 AM, Mars would be just above the horizon.

If he was born at 6:00 AM, Mars would be just past its upper culmination. For our purpose such positions are too vague. In order to apply proper statistical tests we need something more precise, so we must divide the circle of diurnal movement in sectors that will serve as reference points. Astrologers also divide the circle of diurnal movement into sectors, usually twelve, called "houses." Houses are numbered counterclockwise from the Ascendant, and each house is said to correspond to certain life areas whose meanings have been handed down by tradition. But my goal was not astrological but statistical. So I decided to divide the diurnal motion of each planet into 36 sectors starting from the rising point and following the real astronomical movement which is clockwise. I also used larger sectors such as 18 sectors or 12 sectors. The last is astronomically very close to the Placidus houses used by astrologers. In my system, at the moment when a baby enters the world, the sun, moon and planets will each be situated in one or other of the 36 sectors of the celestial dial. If we are studying a group of 1,000 births, we can then count the number of times a planet appears in each sector, just as in gambling we can count the number of times each number appears for 1,000 spins of the roulette wheel. Further details can be found in Appendix 2.

Celebrities and the Planets

By 1951 I had found strange results for successful professionals. One result was especially clear-cut. For 576 distinguished members of the French Academy of Medicine, certain planetary positions appeared at the moment of birth far more than they should have. This result could not be dismissed as mere chance; any statistician would have judged it to be very significant — which is what I did.

My figures showed that these famous doctors seemed to have a peculiar preference for being born when either Mars or Saturn has just risen above the horizon, or when either planet had just passed its upper culmination. By contrast, ordinary people taken at random from electoral rolls showed no such effect.

Somewhat puzzled, I decided to repeat the experiment to see if this strange anomaly would recur. And it did. For 508 different but equally well-known French doctors, their hours of birth showed the same inexplicable tendency to cluster just after the rising and just after the culmination of both Saturn and Mars. Clearly, this needed to be examined in greater detail and, if possible, explained using a sound methodology.

Timing the Destiny

My results encouraged me to gather more data on successful professionals, for example, successful scientists, actors, writers and sports champions. Success may seem like an arbitrary criterion but it is an objective one, because it is relatively easy to agree on whether a person is successful. Success is also a convenient criterion because lists of successful people are readily available in biographical dictionaries.

To my surprise, the more calculations I performed, the more the initial trend was confirmed. On top of that, a more-and-more precise statistical relationship appeared to emerge between the time of birth of great men and their professional success. It was no longer just a matter of the doctors of the French Academy of Medicine. Each professional group seemed to respond to planets in their own special way, especially for the moon, Mars, Jupiter and Saturn. Thus the presence at birth of a planet which has either just risen over the horizon, or just passed its culmination, seemed to "provoke" or "cause" success in some cases, and to "prevent" success in others. The effect was not large (the occurrence of key planets differed typically by only 15% from that expected) but it was statistically highly significant, far too significant to be explained by chance. All of these results have been published several times since 1955, first in my French books and then in my English books (especially my recent *Written in the Stars — The Best of Michel Gauquelin*), so there is no need to give details here. However, for convenience, Table 1 shows the basic findings, and Figure 3 shows how my first findings have been successfully replicated.

At this point, after the gathering of more than 20,000 birthtimes of successful professionals, the conclusion seemed unavoidable: certain planets at the birth of famous people seem related to their destiny. A very challenging and hard-to-explain result!

TABLE 1

The Planets of Success in Different Professional Groups*

after the rise and upper culmination of	*high frequencies of births*	*low frequencies of births*
JUPITER	actors and playwrights politicians military leaders top executives journalists	scientists physicians
SATURN	scientists physicians	actors journalists writers painters
MARS	physicians military leaders sports champions top executives	painters musicians writers
MOON	writers politicians	sports champions

* (from Michel Gauquelin: *Les Hommes & les Astres*, 1960, page 200)

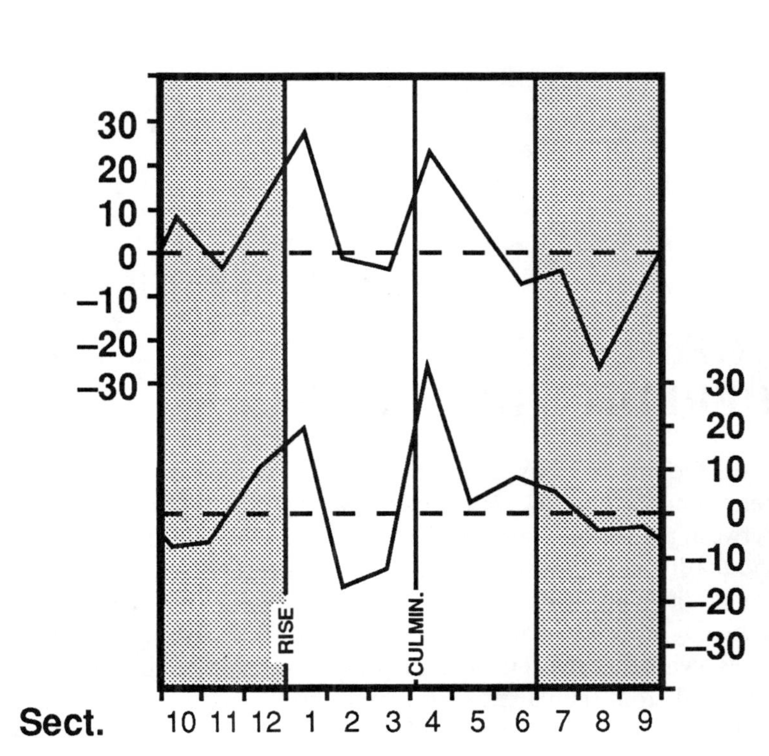

Figure 3: Mars at the Birth of Two Samples of Famous Physicians and Scientists

Upper curve: 1365 French
Lower curve: 1940 European

The two curves show the same pattern: more French and more European physicians were born after the rise and the culmination of Mars.

(The graphs show the difference between observed and expected frequencies: 0 = no difference.)

(from Michel Gauquelin: *Les Hommes & les Astres*, 1960, pp. 68 and 73)

CHAPTER THREE

THE PLANETARY HEREDITY HYPOTHESIS

I was looking for a hypothesis that would let me explain or, at least, better understand the effects under investigation. At first a correlation between planets and famous people seemed impossible to believe. Why should different planets have different influences? Indeed, why should planets have any influence at all? Is a child's destiny and character determined by a few astral rays that mark him or her with an indelible seal? I was aware that the child at birth is physically formed by the potential inherited from the parents. So I could not assert that the planet acts on the chromosomal structure of the child's cells, disrupting and redistributing them to the point of changing his or her mentality. In reality, the same individual born three weeks later would still become a doctor, a sportsman or a painter, according to temperament. This is why I had to believe that the planet adds nothing to the child at birth, but is at most only an "indicator" of a temperamental tendency that will later lead the individual to adopt a career that fits this tendency. For example, consider the champion boxer Muhammad Ali (Cassius Clay). When he was born in Louisville, Kentucky, on January 18, 1942, at 6:15 PM according to his birth certificate, Mars had just culminated. Given the genetic transmission of paternal and maternal characteristics, could one imagine even for a moment that if Mars had been somewhere else, then, Cassius Clay would not have become Muhammad Ali? Inherited disposition and favorable environment made him into the great champion that he became. If a planet is to be integrated into his circuit, then due account must be taken of heredity and environment, of which heredity seems the easiest to cope with.

In 1959 I was therefore ready to consider that the diurnal planetary positions at the birth of a child might be the sign of some heredity factor. Now, if the planets really are linked to heredity, then parents and children should display a tendency to be born under similar planetary patterns. So I needed to gather samples of parents born when Mars was rising or culminating, for instance, to see if their children tended to be born with Mars in the same position. More generally, the reality of a planetary heredity would be demonstrated if I could show statistically that children tend to be born under the same cosmic conditions that prevailed at the birth of their parents.

Birth Data of Parents and Children

The checking of my planetary hypothesis raised a lot of methodological problems that required much careful work. In a controversial field like this one, an impeccable methodology is essential which is why the appendices of this book are devoted to it. Here I will describe only the steps necessary for testing the hypothesis.

First I needed to know the date, time and location of each birth so that planetary positions could be determined. Once again, as for my famous professionals, I obtained the birth data from registry offices. Birth records are objective information for both parents and children.

Altogether I gathered, with the help of my wife (now ex-wife) Françoise, 24,961 birth data, allowing 16,037 parent-child comparisons for each celestial body. The births spread over a century, from January 1, 1850, for the oldest parents to December 31, 1945, for the youngest children. All information received from the registry offices or copied by our own hands is kept in my laboratory, and any person who wants to check any data is welcome to do so.[1]

I did not content myself with looking only at Mars, Jupiter, Saturn and the moon, the only four that showed significant results with famous people. I hoped that other celestial bodies might produce something significant with this new hypothesis. Venus, the planet closest to the Earth, seemed particularly surprising because no results were obtained for professional groups. Accordingly, for every one of the 24,961 births I determined the position of the sun, moon and planets in the 36 sectors of diurnal movement. Figure 4 shows one of the heredity

1. Since this was written, all the birth and planetary data of the heredity experiment (M. & F. Gauquelin, *Series B*, Volumes 1-6, 1970-71) have been published by my laboratory.

index cards bearing the information necessary for statistical analysis of the results. All the parent-child data were put on index cards like this.[2]

L.	Jean	21.2.1896,0h	Fontaine (Yonne)
P.	Angéle	7.11.1890,21h	Iffendic (I & V)
	fem.	6.2.1924,0h30	Créteil (Seine)

	SO	MO	ME	VE	MA	JU	SA	UR	NE	PL
Father	27	17	29	30	30	12	36	36	16	16
Mother	24	29	25	22	17	17	29	25	5	5
child	28	27	30	25	33	33	2	25	10	13

Figure 4: A Planetary Heredity Index Card

At the top is the birth information for the father, mother and child. Below are the sector positions for the ten celestial bodies of the solar system. Their names are abbreviated. This document allows sector positions to be compared between the child and each parent.

Astronomical Conditions of the Hypothesis

The astronomical conditions of the planetary heredity hypothesis are summarized in Figure 5. We can see how it works by referring to the case shown in Figure 4. This shows that at the birth of Jean L., the father, Jupiter was in sector 12. This sector is in the significant culmination zone (the sectors included in the rising and culmination zones are shown in Figure 5; further details can be found in Appendix 3). However, when the child was born, Jupiter was in sector 33, which is not in a rising zone or culmination zone. So according to the hypothesis there is no heredity link in this case between father and child for

2. Needless to say, this was before the age of computers so that all work had to be carried out by hand. Now all data are stored in the Astro Computing Services' computer, and all calculations are done by machine.

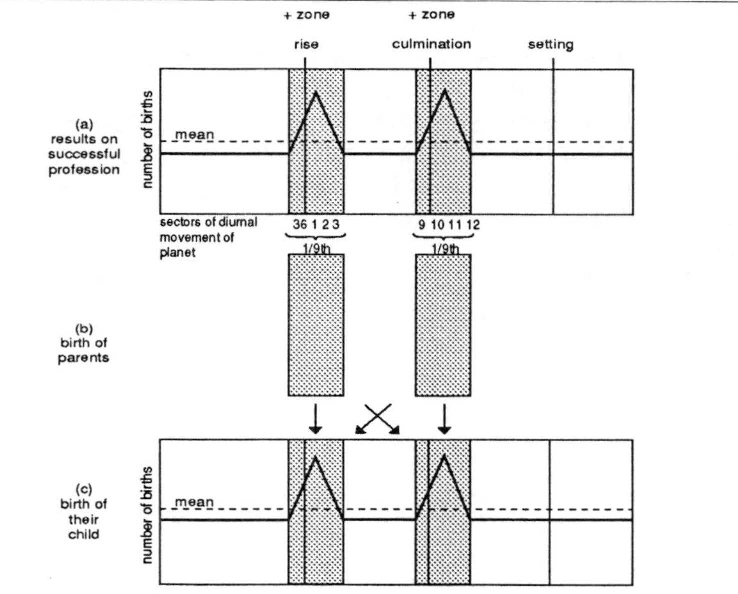

Figure 5: Planetary Effect of Heredity: Astronomical Conditions of the Hypothesis

(a) The results for successful professionals showed that the celestial body had one effect when in (+) zones, the rising and culmination zones (shaded), and the opposite effect when elsewhere. If the difference is in some way linked to genetic differences, then the effect should be linked to heredity (the planetary heredity hypothesis) and should show up between parent and child.

(b) Therefore, if the planet is in a shaded (+) zone at the birth of the parent...

(c) ...then at the birth of their child, the planet should be more often in a shaded (+) zone than elsewhere.

And vice versa: if the planet is elsewhere at the birth of the parent, then at the birth of the child it should be more often elsewhere than in shaded zones.

The arrows between the shaded zones of (b) and (c), and which which appear in subsequent graphs of this type, indicate that the two zones are equivalent. That is, the hypothesis links the left zone of (b) to the left **and** right zones of (c), and similarly for the right zone of (b). In other words, planets rising or culminating for children counted as a hit, whether a parent had a planet only rising or only culminating.

Jupiter. Saturn, on the other hand, is in sector 36 (rising zone) for the father and in sector 2 (rising zone also) for the child. So, for Saturn, there is a heredity link between father and child.

In the course of this book, the graphs will follow the style of Figure 5. [3] But, of course, graphs are not sufficient evidence for planetary heredity. A statistical analysis is needed to tell us if the observations are significant. In my studies I used the chi-square test, a well-known statistical test that is described in detail in Appendix 3. The chi-square test indicates whether the result is significant (in which case the planetary heredity hypothesis is supported) or not significant (in which case the hypothesis is not supported).

3. Note that in Figure 5, each shaded zone consists of **four** out of the 36 sectors. In my heredity experiments the planetary conditions of parents are always measured using the two sets of 4 sectors (out of 36).

CHAPTER FOUR

PLANETARY HEREDITY RESULTS

The planetary heredity hypothesis has been tested for the ten celestial bodies of the solar system. The birth data allow 16,037 parent-child comparisons. The chi-square test showed that the results were significant for some planets but not for others.

The Distance Factor

The significant group consists of the moon, Venus, Mars, Jupiter and Saturn. These are the four celestial bodies that gave significant results for famous people, and Venus. Their results support the heredity hypothesis. The statistical values are:

	chi-square	probability
Moon	9.40	.002
Venus	9.80	.002
Mars	14.59	.0001
Jupiter	3.02	.08
Saturn	4.80	.03

The results for the other celestial bodies did not support the heredity hypothesis:[1]

	chi-square	probability
Mercury	0.14	.71 (not significant)
Uranus	0.54	.46 (not significant)
Neptune	1.90	.17 (not significant)
Pluto	0.00	.99 (not significant)

1. No heredity effect was found for the sun. That result will be analyzed separately (Chapter Nine).

These results are summarized in Figure 6. It is interesting to note that the significant planets are all visible to the naked eye, whereas the nonsignificant Mercury, Uranus, Neptune and Pluto are generally not visible. Thus planetary heredity may be partly related in some way to visibility, or perhaps to distance modified by planetary mass. This observation, however, has to be confirmed by the replication of new experiments.

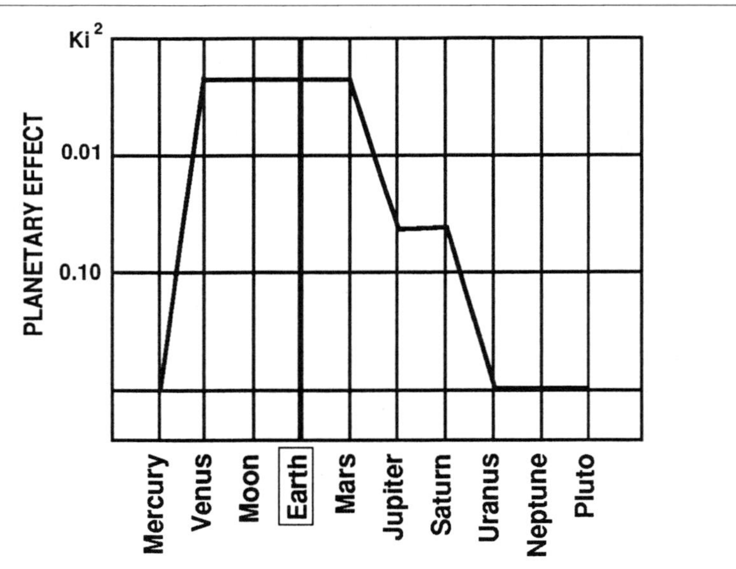

Figure 6: The Planetary Effect and the Planet's Distance from Earth

Abscissa: the planets on each side of Earth according to their distance from it. Planets at the left of Earth are close to the sun; at the right, they are far from the sun. (The figure does **not** respect accurate distance scale.)

Ordinate: the level of probability of result by chi-square test for each planet (the accurate values of chi-square test are given in the text).

More details of the heredity effect are given in Figure 7. The moon, Venus, Mars, Jupiter and Saturn distributions are remarkably similar and look very much like the hypothetical curve given in Figure 5. Figure 7 shows how children tend to be born when the planet rises or culminates if it was in the same position at the birth of their parents, and vice versa. Five times, peaks of frequency occur in the same sectors of the diurnal movement, which are precisely the zones predicted by the hypothesis. Such a result is quite unlikely to be due to chance.

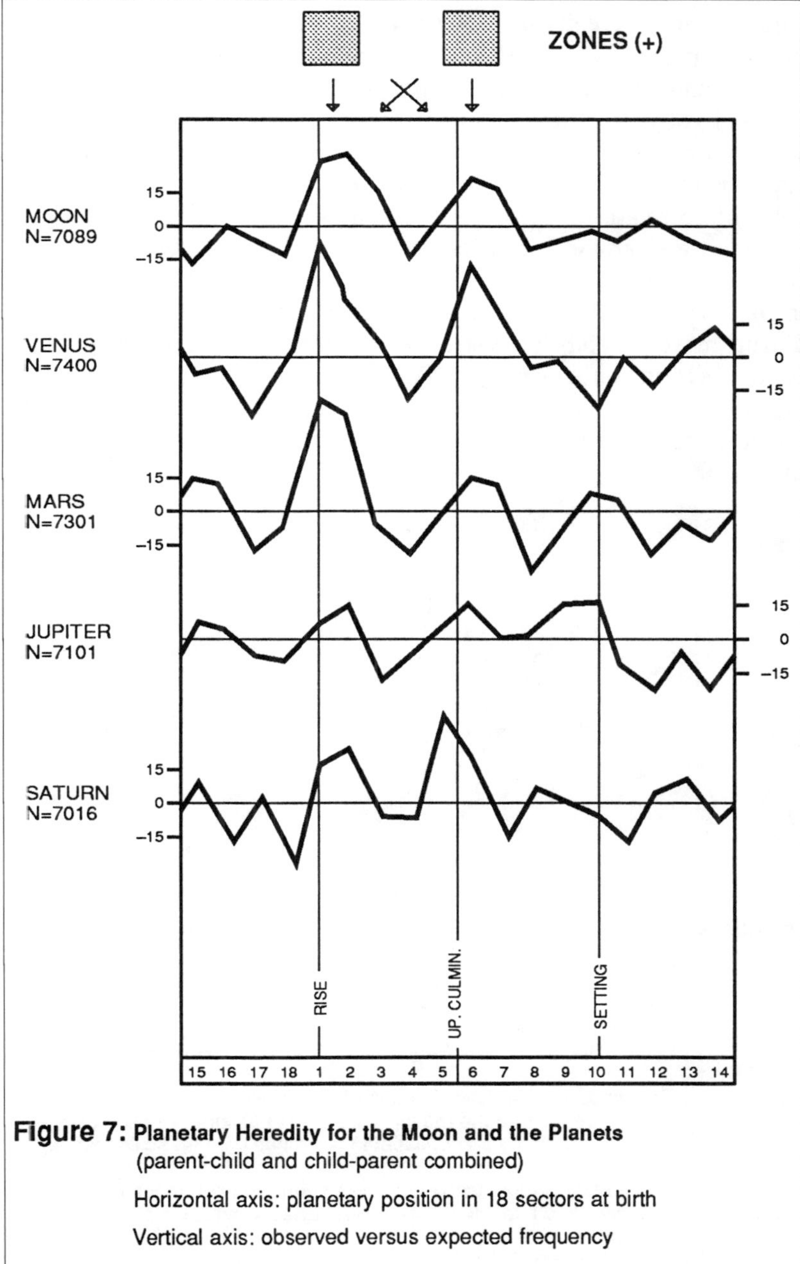

Figure 7: Planetary Heredity for the Moon and the Planets
(parent-child and child-parent combined)

Horizontal axis: planetary position in 18 sectors at birth

Vertical axis: observed versus expected frequency

Profession and Heredity

The planetary heredity hypothesis was formulated in order to test the possibility that heredity effects might help to explain the results previously obtained with famous people. The present results enhance this possibility. In Figure 8 the combined observations for successful professionals are compared to the combined observations from the heredity experiment. The two distributions are remarkably similar, the coefficient of correlation being +.83, probability .00002.

Figure 8 shows the profession effect is nearly twice as strong as the heredity effect. It also shows that planets have an effect not only in the rising and the culmination zones but also in other parts of the diurnal movement albeit less strongly. In particular, the setting and lower culmination zones seem to weakly echo the effects found for the rising and upper culmination zones. Thus every six hours, when the planet has just crossed the horizon or meridian, there is a maximum of frequencies.

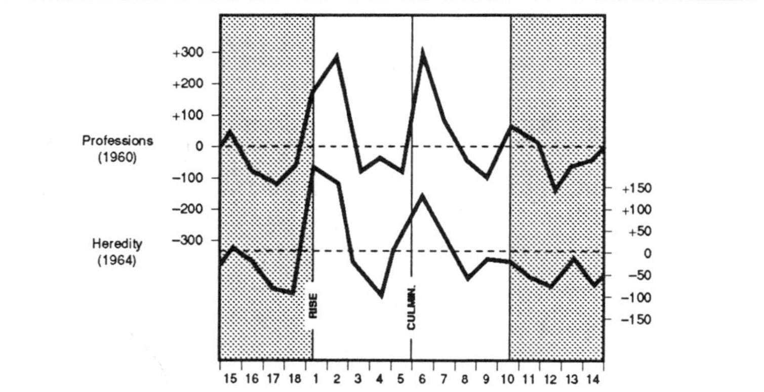

Figure 8: Identical Pattern of the Planetary Effect between Profession and Heredity

Top curve: Profession. Combined distribution for Mars, Jupiter, Saturn and the moon at the birth of successful professionals (cf. Michel Gauquelin, *Les Hommes & les Astres*, 1960, pp.195).

Bottom curve: Heredity. Combined distribution for Mars, Jupiter, Saturn, Venus and the moon at the birth of children whose parents were born with the same planet in (+) zones (rise and culmination).

For the two curves, the vertical axis shows the difference between observed and expected frequencies. Shaded area = planet below horizon.

The similarity between the profession and heredity curves may lead to interesting conclusions. After my work on famous people, I wrote: "There is a correlation between Mars, Jupiter, Saturn, and the moon and the birth of well-known individuals." I may now add: the time of birth of these well-known individuals is itself related to their heredity. In other words, the same effect is related to both planetary heredity **and** the general orientation taken by human beings in the course of their lives. This suggests that a genetic sensitivity to cosmic factors at birth exists for every man and woman, and not just for successful people, in which case the planetary positions at birth will tend to reveal (albeit imperfectly) the constitutional type of the child. (Replications of the heredity experiment are described in the Postscript.)

CHAPTER FIVE

CHILDBIRTH CONDITIONS

Do the childbirth conditions I described in Chapter One have an effect on planetary heredity? According to my observations, normal individual variations do not seem to play a significant role. For instance, although primiparous deliveries last about three hours more than multiparous deliveries, there is no significant difference between them with regard to planetary heredity: the primiparous sample consists of 1,027 deliveries (average duration of labor: eight hours); the multiparous sample consists of 1,671 deliveries (average duration of labor: five hours). The number of similarities between parent and child when the parent has a planet in rising or culmination zones is: primiparous, expected frequency 285, observed frequency 306, difference +21 or 7%; multiparous, expected frequency 449, observed frequency 474, difference +25, or 6%. Both differences are positive but there is virtually no difference between them.

Factors like age of the mother, weight of the child and time of onset of labor do not seem to play a role either. However, my samples may have been too small to detect any effect, so these experiments should be replicated on a larger scale before attempting to draw conclusions.

Surgical Intervention

We carefully noted any births due to surgical intervention such as cesarean section or the use of forceps. The 24-hour distribution of these medical births is completely different from the daily rhythm of natural deliveries. Figure 9A shows the medical curve previously described in Chapter One. Medical deliveries tend to occur more often during the middle of the day when, under normal conditions, the frequency of births is low. In this sample the planetary heredity disappears completely and even tends to be reversed (Figure 9B); 288 parent-child pairs with medical births were gathered, giving for the five significant planets a total of 1,440 planetary comparisons. The number of similarities between parent and child when the parent has a planet in rising or culmination zones is 74, expected frequency 89, difference −15, opposite to that observed with normal births. However the sample is too small for any definitive conclusion to be drawn. Replications are needed (see Postscript).

Drugs during the Delivery

As I have previously described, natural deliveries have been modified by various drugs from simple painkillers to labor inducers like oxytocin, which causes the uterus to contract. Unfortunately, medical information concerning the delivery is usually not available. We can, however, use a valid general indicator, namely, the 24-hour distribution of births. If the distribution follows the natural curve, we may expect a clear planetary effect. But if the natural pattern is not there, it means that a large proportion of births have been induced or accelerated by physicians and midwives (the more distorted is the natural pattern, the larger is the proportion of medical births in the sample), in which case we may expect at best only a weak planetary effect.

This assumption has been verified using a sample of births from the suburbs of Paris. The births occurring after the end of 1937 were mostly taken from the maternity ward of a large hospital recently built at the time. Soon the 24-hour pattern of birth was changing, a clear indication of the introduction of new obstetrical policies. The natural rhythm that is visible up to 1937 disappears almost completely for the 1938-45 period; and so does the planetary heredity. So I divided the sample into three groups of about the same size, keeping the year 1938 as a turning point:

1) Children born between 1923 and 1931 in Paris.[1]
2) Children born before 1938 in the suburbs.
3) Children born after 1938 in the suburbs.

1. The last year of our observation in the city of Paris is 1931.

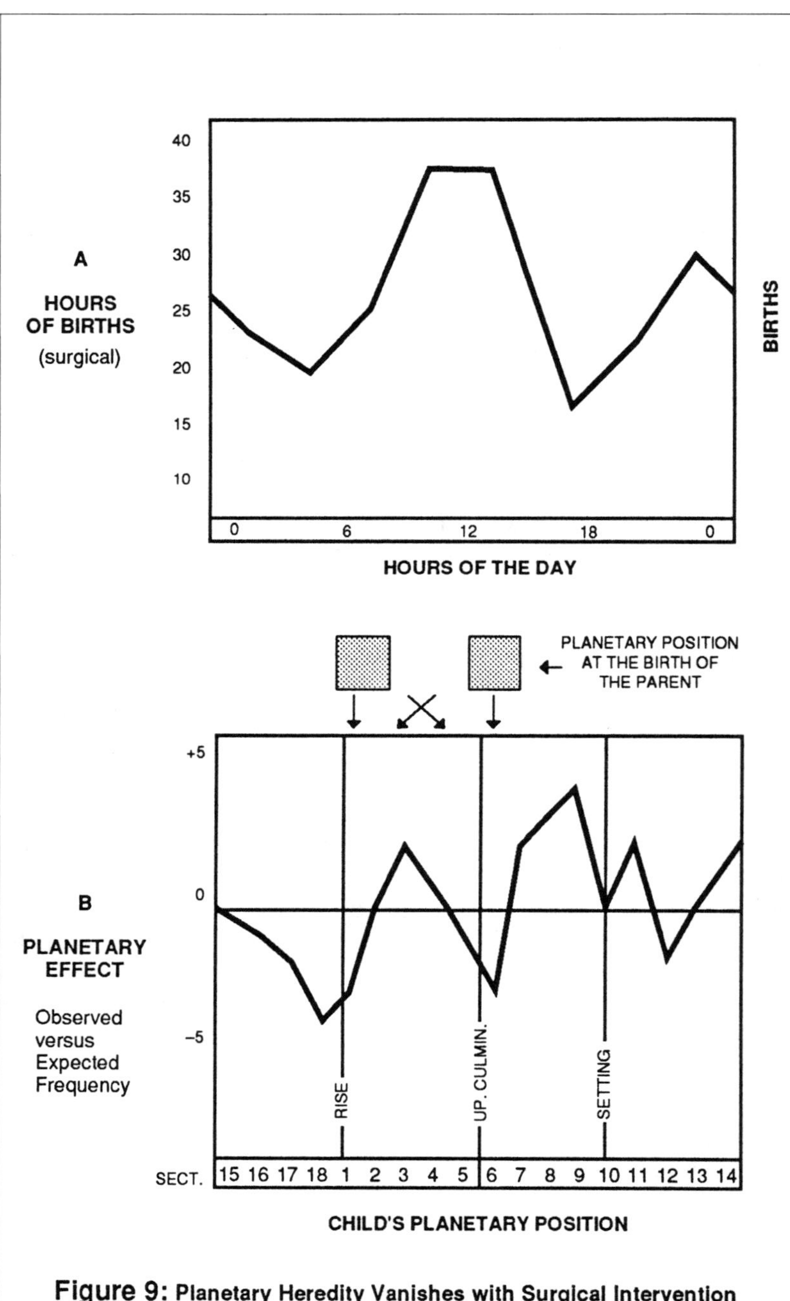

Figure 9: Planetary Heredity Vanishes with Surgical Intervention

The chi-square values for these three groups are:[2]

1) Paris up to 1931: chi-square 26.03, probability = .0001
2) Suburbs before 1938: chi-square 18.93, probability = .002
3) Suburbs after 1938: chi-square 8.52, probability = .13

In the last sub-group, the observed frequency of parent-child planetary similarities is still above the expected frequency but the difference is not significant. These results show that planetary heredity has strongly decreased after 1938 — at least for the sample under investigation.

Shift of the Effect in Recent Years

Figure 10B shows that the 18-sector distribution **after 1938** is shifted about one sector in advance of the 18-sector distribution observed **before 1938**. The two curves have similar periodicities. But the peaks of the recent one always precede the peaks of the older one, hence the chi-square value (which is tied to the rising and culmination zones) is reduced. The planetary positions for children born after 1938 are in effect "dislocated" from those of their parents born with the planets in rising or culmination zones. Such dislocation is not surprising. We know that modern medical techniques tend to shorten the delivery, so that births after 1938 occur 1-2 hours earlier in the day than births before 1938 (Figure 10A).

A special inquiry made at the maternity ward of Créteil Hospital, where most of our "after 1938" children were born, produced interesting results. We noticed that a relaxing drug called "Spasmalgine" was administered fairly systematically to mothers who arrived at the hospital after the onset of labor. About "Spasmalgine," Prof Vignes, an obstetrician, writes: "This medication is widely used today, and it is correct to say that it favorably helps the delivery by accelerating and regulating the uterine contractions."[3] This supports the view that dislocation of the heredity effect after 1938 is due to the use of drugs like Spasmalgine which reduce the duration of labor. One might argue that no such dislocation was observed for the 3-hour average difference between primiparous and multiparous deliveries, so no firm conclusion is possible. However, the two experiments cannot be compared since

2. Here the chi-squares are calculated with five degrees of freedom (see Appendix 3). This test takes into consideration the five values of the chi-square test computed for each of the five significant planets.
3. H. Vignes, *Les Douleurs de l'Accouchement*, Paris, 1951, pp. 162-163.

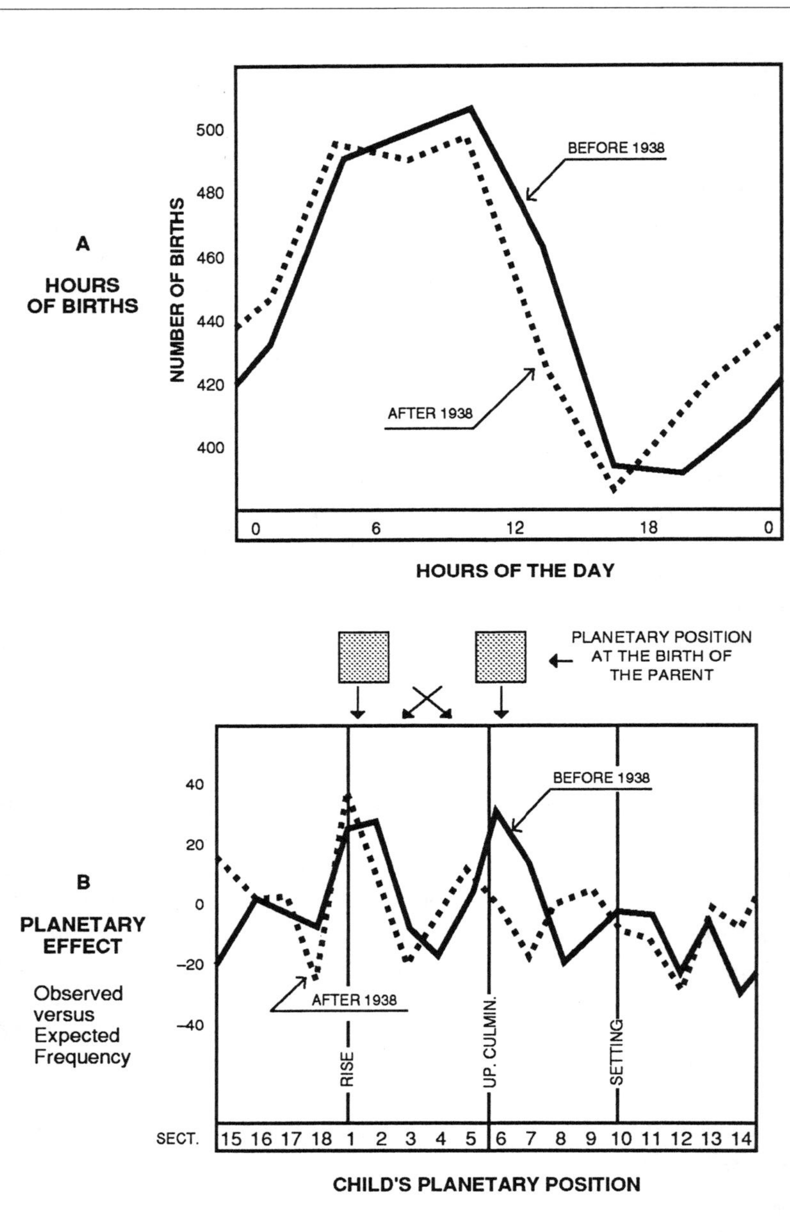

Figure 10: Diurnal Rhythm and Planetary Effect Shift Ahead after 1938

primiparous/multiparous comparisons were made on normal deliveries, not on induced ones. So the meaning of the two experiments is rather different.

Irreversible Evolution

As yet, we do not know what biological mechanisms the planets may affect during delivery, but they are likely to operate at a very subtle level. Drugs and surgical interventions modify the connection between mother and fetus so that the hormonal role of the fetus is perturbed or even annihilated. For us this is an unfortunate situation since it is only through the reaction of the fetus that planetary heredity can be observed.

In our experiment on heredity, we did not gather children born after 1945. But, since then there has been a tremendous increase in the proportion of induced births or births by cesarean section, including those performed for "convenience."[4] Any experiments on planetary heredity that use births from 1950 onward should include a careful analysis of each case in order to exclude non-spontaneous births. This is an absolute necessity for anyone who aims to replicate the results presented in this book.

4. All of this has been recently described in detail in Chapter 9, "Neo-astrology under Attack," of my book *The Truth about Astrology*, Oxford, Blackwell, 1983 (published in the U.S. as *Birthtimes*, New York, Hill & Wang).

CHAPTER SIX

PLANETARY HEREDITY IN THE FAMILY

Is planetary heredity affected by the sex of parent and child? Is it stronger if both parents have the same planet in rising or culmination zones? Is it affected by birth order? This chapter will try to answer these questions.

Effect of Sex

In genetics, some factors are known to be sex-specific. What about planetary heredity? To find out, I divided my material into four categories: father and son, father and daughter, mother and son, mother and daughter. Then, for each category, I looked at the result for the five significant planets combined. I found the following chi-square values with five degrees of freedom:

— Fathers (whatever the sex of the child): 15.37, p = .009, very significant.
— Mothers (whatever the sex of the child): 32.84, p = .000004, very significant.
— Sons (whatever the sex of the parent): 23.24, p = .0003, very significant.
— Daughters (whatever the sex of the parent): 20.51, p = .001, very significant.

Each chi-square value is very significant showing that planetary heredity is not sex-specific. The mother group gives a higher value of chi-square than the father group because the data includes more mothers than fathers (see Appendix 1 for an explanation), so we cannot conclude that mothers show a stronger effect than fathers. The pattern of the effect for each category is shown in Figures 11 and 12. The patterns are all very similar, which confirms that planetary heredity is not sex-specific.

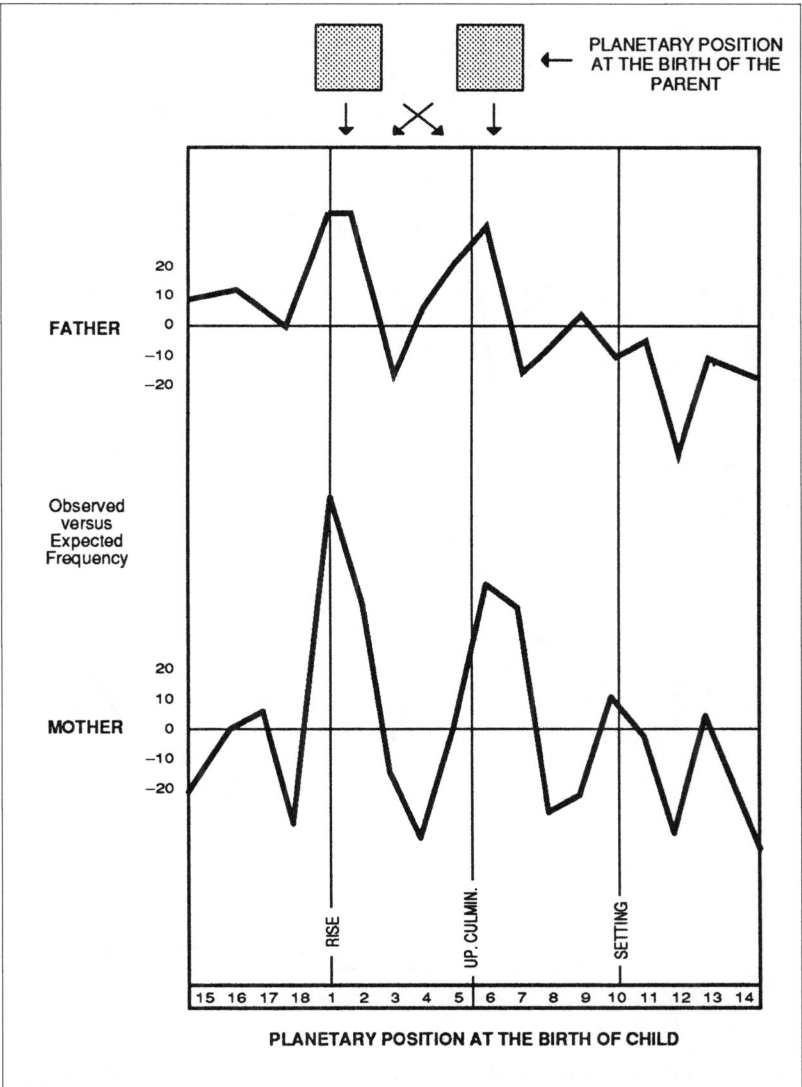

Figure 11: With Father and Mother the Planetary Effect Shows the Same Pattern

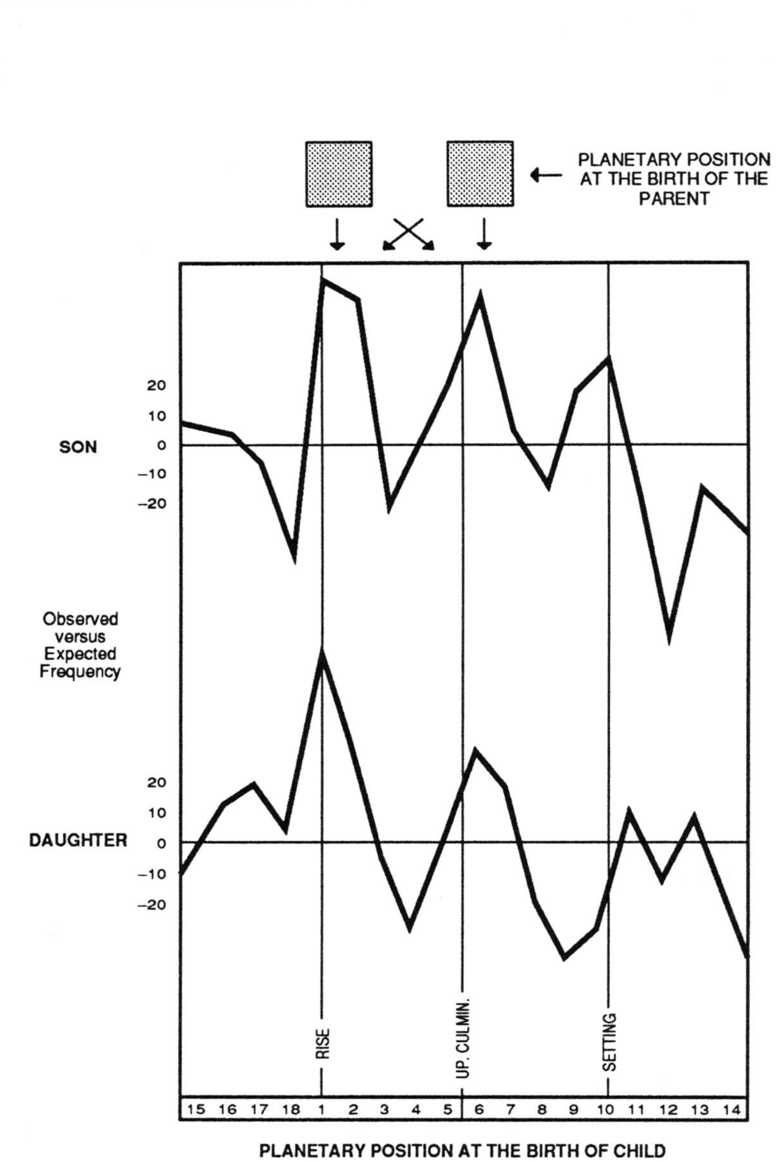

Figure 12: With Son and Daughter the Planetary Effect Shows the Same Pattern

The existence of an effect between father and child is fairly good evidence that the fetus helps to start its own birth. If it did not, so that only the mother started the birth, then no effect would be observed, simply because the father cannot play any physiological part in the delivery. Therefore it seems that, when the time to be born has arrived, the fetus tends to react to cosmic factors according to some constitutional elements inherited from father and mother. It is very encouraging that medical evidence (briefly mentioned in Chapter One) supports the role of the fetus during delivery. Otherwise planetary heredity between father and child would have been very puzzling.

Plus-plus Parents vs Minus-minus Parents

So far, I have presented only the results for planetary heredity between child and **one** parent. But it is important to look at planetary heredity between child and **both** parents together. So the 3,487 cases which had birth data for both parents were analyzed in three categories:

(+ +): Both parents born with the same planet in the rise-culmination zones.
(+ −): One parent born with the planet in the rise-culmination zones, the other parent born with the planet elsewhere.
(− −): Both parents born with the planet outside the rise-culmination zones.

The results for the five significant planets combined are:

(+ +) Parents: +12%	Difference between observed and expected
(+ −) Parents: +7%	frequency in rise-culmination zones at
(− −) Parents: −6%	birth of child, as % of expected
	frequency.

The results suggest that the heredity effect is nearly doubled when both parents are born under the same planetary positions. To check this result I analyzed the large number of cases where, due to the numerous difficulties we met (see Appendix 1), we had managed to obtain the birth data of only one parent. I examined two categories:

(+ ?): One parent (father or mother) born with the planet in the rise-culmination zones; details unknown for the other, hence the question mark.

(– ?): One parent (father or mother) born with the planet outside the rise-culmination zones; details unknown for the other, hence the question mark.

The results for the five significant planets combined are:

(+ ?) Parents: +7%	Difference between observed and expected frequency at birth of child, as above.
(– ?) Parents: –2%	

When compared to the previous results, the heredity effect seems weaker when there is incomplete information about parents. This is easy to explain because on the average, 7/9ths (Figure 5) of the unknown parents will have the planet outside the rise-culmination zones, which of course will dilute the heredity effect in the (+ ?) case. And vice versa for the (– ?) case.

The two sets of results are consistent, and show a consistent variation with parent condition. When the results are summarized as below, their consistency is readily apparent:

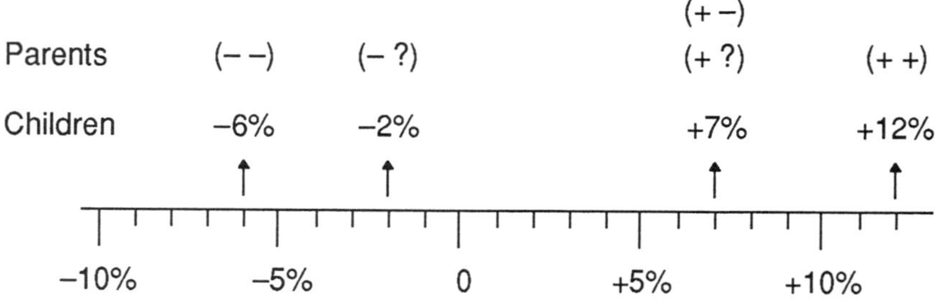

The pattern of the effect for each category is shown in Figure 13; the results are consistent with the above observations. Figure 13 shows that the heredity effect is stronger for children with (+ +) or (– –) parents than for children with (+ –) or (+ ?) parents. The curves clearly show the weakening that occurs when the birth data is known for only one parent. Unfortunately, due to the pressure of material constraints, I often had to work with such incomplete data. However this weakening is also a good indication that the results are not fluctuating at random but are following some physical logic.

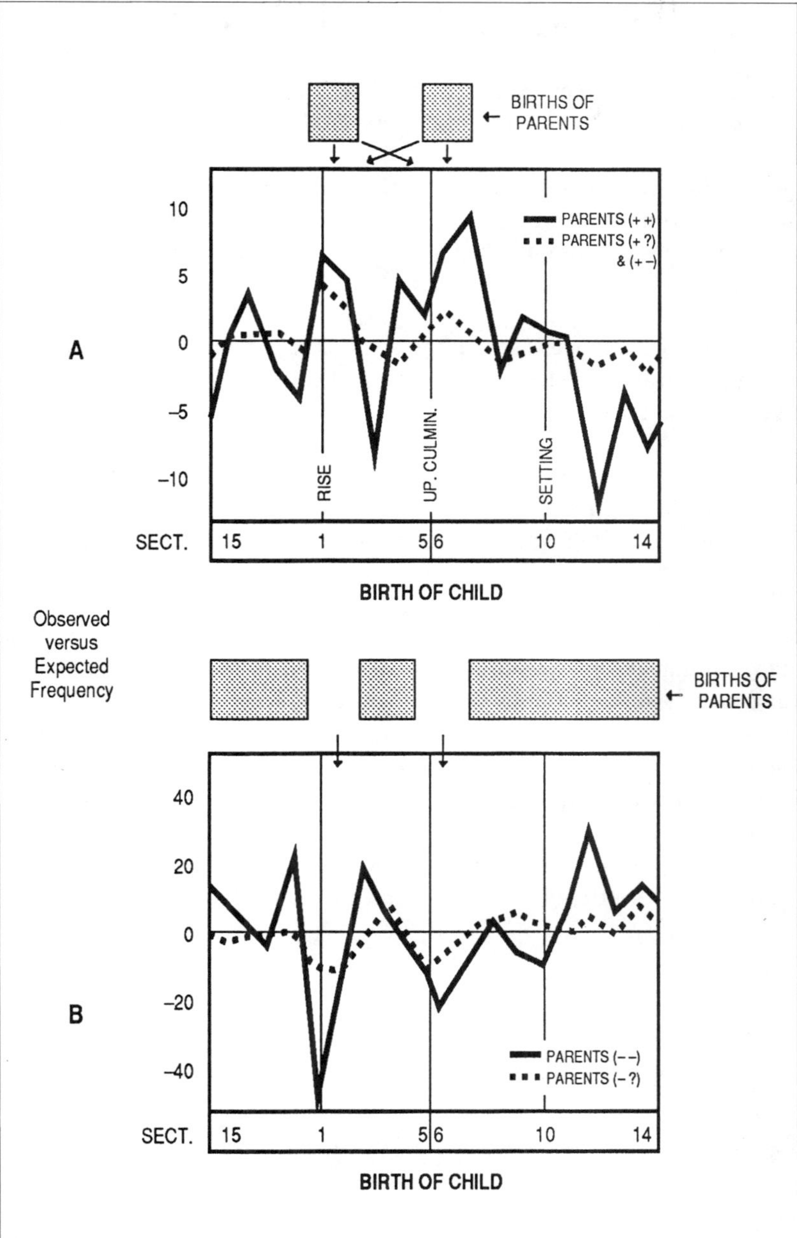

Figure 13: Parents with Same Planetary Heredity: The Planetary Effect Increases for Children

Siblings: Effect of Birth Order

Is the heredity effect the same for all siblings? To find out, the 6,691 cases with more than one child were analyzed in two categories:

— Siblings with parents born with the planet in rise-culmination zones, i.e., from (+) parents.
— Siblings with parents born with the planet outside rise-culmination zones, i.e., from (–) parents.

For each planet, I made the same comparisons between siblings as I previously made between parents and children. The chi-square value with five degrees of freedom are:

— Between siblings with (+) parents: chi-square (5) 56.06, p = .0000001, significant;
— Between siblings with (–) parents: chi-square (5) 12.13, p = .03, significant.

The results show that siblings with (+) parents have more planetary similarities in rise-culmination zones than are expected by chance; and vice versa for siblings with (–) parents, but to a lesser degree. We may therefore assume that, in a family of several children, planetary heredity is not affected by birth order. It is worth stressing that these results are a consequence of the heredity effect between parents and children; that is, we are not entitled to postulate the existence of a specific "siblings" effect, independent of that observed between parents and children. More details can be found in Appendix 3.

CHAPTER SEVEN
OTHER TIMES OF THE LIFE

It would be very surprising if planetary effects could be observed only at the time of birth. Similarly, there is no good reason to work only with humans. It would be very interesting to look for planetary heredity among other animals, especially in mammals such as ewes, monkeys and horses. Even negative results would be full of information.

So I looked at the other important times, namely, the onset of labor, and death. The results presented here are not conclusive and should be considered only as invitation to further study.

The Onset of Labor

In Chapter One, we saw how the time of birth is determined by the physiological processes of the labor that precedes it. We also saw that the onset of labor shows a very clear diurnal rhythm. So is the onset of labor related to planetary heredity? According to my hypothesis that planets may trigger birth, there should be a stronger heredity effect at the beginning of labor, i.e., the trigger time, than at the end, i.e., birth time.

Fortunately, for the birth data between 1937 and 1945 already analyzed (N = 3794), the time of onset of labor was available from hospital records. So, after much work we were able to compare planetary positions at the birth of the parent with planetary positions at the onset of labor. No comparison was possible with planetary positions at the onset of labor when the parent was born, because this information was not available. So, I was left with a rather awkward comparison between a birth time and, a generation later, an onset of labor time.

The results show a heredity effect similar to that previously observed, but to a much lesser degree, the chi-square value for all five significant planets combined being only 2.90, probability .09. As before, the effect was increased when both parents were born with the same planet in rise- culmination zones. The diurnal pattern shows the familiar peaks after rise and culmination; but a third peak appears after set (Figure 14). This study suggests that a heredity effect may exist at the time of onset as well as at the birth time, but until the results are replicated, no conclusion is possible. Because the onset times were imprecise, being recorded only to the nearest hour or two, a replication is urgently needed, hopefully with more accurate data (a replication is described in the Postscript).

If an effect at the onset of labor was confirmed, it follows logically that planets could affect the fetus throughout the delivery, either continuously or discontinuously. Thus the time of birth would lose its privileged position. But for the moment this is only vague conjecture.

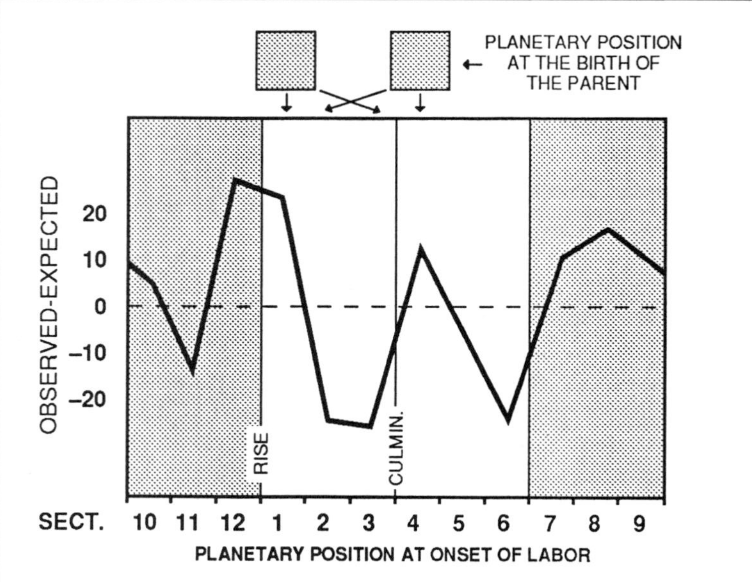

Figure 14: Planetary Heredity and Onset of Labor

When labor begins, planets tend to be in rise-culmination zones if the parents (father or mother) were born under the same conditions.

The graph shows the difference between observed and expected frequencies for the five significant planets combined. Further details are given in Appendix 4. In this case 12 sectors are used instead of 18 because the sample is rather small and the statistical tendencies are rather weak.

The Time of Death

The genetic structure of a human being remains constant during life, as does the daily rotation of the planets. So the planets have a lifetime in which to consolidate their influence. Could they have any effect on death?

The time of death, like the time of birth, has been officially recorded in France for over a hundred and fifty years. So ample data exists for a statistical inquiry on death.

Death is not independent of external factors. There is a 24-hour rhythm of death. Thus natural deaths occur more often in the early morning than in the afternoon, which is similar to the daily rhythm of birth but is less marked. Cosmic phenomena like solar flares and magnetic storms can provoke fatal complication among people already in a critical condition. So a planetary influence on death is not inconceivable.

The question is: does a person tend to die under the same planetary conditions as existed at birth? To find out, a sample of 4,779 birth-death comparisons was gathered. The births and deaths occurred in three suburbs of Paris and do not represent a random sample because of practical constraints (see Appendix 1). But this is not a serious inconvenience for this kind of study. Results show that, for four planets out of five, the answer is yes. In these cases there are more deaths than expected when the planet is in the same position as at birth. However, the effect is very weak, and the chi-square value is significant only for Venus, which incidentally is quite the opposite of what astrologers would predict. For the five significant planets combined, the chi-square value is 2.74, probability .10, or marginally significant. Figure 15 shows the diurnal pattern at death for people born with planets in the rise-culmination zones. Once again the familiar peaks after rise and culmination can be seen, but because the effect is so small, this result is only suggestive, needless to say.

It would be interesting to replicate the study with more refined data. The ways to die are numerous. It is therefore important to know the cause of death. For our purpose, violent death, unfortunately not an exceptional event, is not natural and should be discarded. Death involving drugs or medication should be discarded for the same reason. Natural deaths should be accepted as such only after a careful check of each case. Sudden natural deaths, such as fatal heart attacks, might be the best to study first. Unfortunately, in France at least, the cause of death is not recorded on the death certificate, so we have to search the records of the hospital where death occurred. This complicates the gathering of significant data.

I hope that other research workers will be interested enough to replicate this study in the years to come.

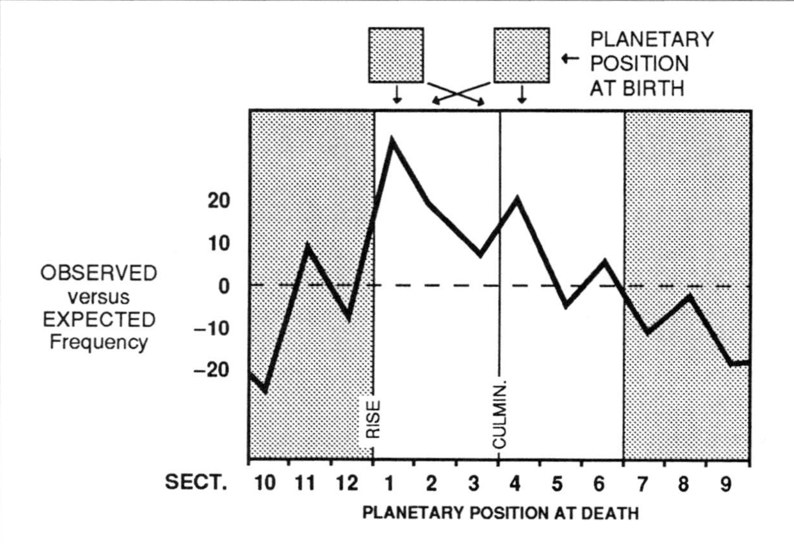

Figure 15: Planetary Effect between Birth and Death

The curve shows that people with planets in rise or culmination zones tend to die with planets in the same position more often than expected.

The graph shows the difference between observed and expected frequencies for the five significant planets combined. Further details are given in Appendix 4.

In this case 12 sectors are used instead of 18 because the sample is rather small and the statistical tendencies are rather weak.

CHAPTER EIGHT

ASTROLOGY AND HEREDITY[1]

So far I have written about planetary heredity as if astrological doctrine does not exist. This is because I have preferred to examine cosmic influences as a science. It is now time to get back to astrological doctrine.

The question of heredity has been important to astrologers since Kepler wrote: "There is one perfectly clear argument beyond all others in favor of the authenticity of astrology: this is the common horoscopic relation between parents and children" (*Harmonices Mundi*). Kepler further explained his thoughts in a letter written in 1598 to Maestlin, his skeptical master: "Look at the relationship between births. You were born under a conjunction of the sun and Mercury and so was your son, Mercury being behind the sun at the moment of both your births. You were born at the moon-Saturn trine; your son was born almost at the moon-Saturn sextile."

At the beginning of the twentieth century, the French astrologer Paul Choisnard (1867-1930) announced that he had statistically proven the laws of astral heredity: "The birth charts of blood relatives demonstrate a more frequent similarity than the charts of unrelated persons" (1919). Choisnard's law implies that there is a strong tendency for children to be born with the Sun, Moon, Ascendant and MC in the same zodiacal position as their parents.

Swiss astrologer Karl E. Krafft (1900-1945) subsequently put

---

1. This chapter is a translation of: Michel Gauquelin, "L'hérédité astrale" published in *Les Cahiers Astrologiques*, 1962, pp. 135-143. It was not included in the 1966 French edition of *Planetary Heredity*. It was partly published privately in English by my laboratory in 1978 (*Series D*, Volume 3, Part 2). I think the material is of particular interest for astrologers, so I include it here.

emphasis on some of Choisnard's statements in his *Traité d'Astrobiologie* (*Astrobiological Treatise*) published in 1939. This led astrologers to believe that zodiacal heredity had been proven by statistical studies. In 1960, for instance, the well-known Austrian astrologer Wilhelm Knappich stated: "Like Kepler, we consider as heredity factors the Sun, Moon, Ascendant and MC in the Zodiac, the Houses and the Aspects."

In *L'Influence des Astres* (*The Influence of the Stars*), 1955, I put Choisnard and Krafft's claims to the test using large samples. My results showed that their astrological laws about heredity were based on faulty procedures. I made a larger and more detailed investigation several years later with results that appeared in *Les Cahiers Astrologiques* of May/June 1962. A detailed account of my 1962 report follows.

Heredity in Signs of the Zodiac

Is there a tendency for children to be born under the same zodiacal sign as their parents? To find out, I conducted experiments on the zodiacal signs that are said to be the most important, that is, the Sun, Moon, and MC. The Ascendant was not included because its expectancy is too variable for a reliable test. Data on several thousand births were collected. (In fact, it was the same data used for the first planetary heredity investigation carried out in Paris, see Appendix 1.) Altogether there were 7,846 births, that is, 3,923 parent-child pairs.

The results are shown as observed and expected frequencies in Table 2.

For instance, 39, the first figure of the table, indicates that the Sun was in Aries 39 times (parent **and** child). The second figure, 35, is the number expected by chance. Totals at the bottom of each column show that the Sun, for instance, was expected 335 times in the same sign for parent and child, and was actually observed 330 times.

The results show no trace of zodiacal heredity for the Sun, Moon and MC. The chi-square values for observed versus expected frequencies are not significant in any of the cases, whether taken individually or in combination. The overall total of 977 is less than the expected total of 991, so, overall, the results are not even in the right direction. No difference was found between fathers and mothers.

TABLE 2

Heredity in Zodiacal Signs*

Number of cases where parent and child have the Sun (or Moon or MC) in the same sign. Based on 3,923 comparisons for each astrological factor.

Sign	Sun		Moon		MC		Total	
	Obs	Exp	Obs	Exp	Obs	Exp	Obs	Exp
Aries	39	35	38	29	22	24	101	92
Taurus	28	28	19	26	29	29	76	83
Gemini	18	29	34	31	22	28	73	84
Cancer	20	28	39	32	23	30	80	86
Leo	26	26	30	29	28	24	84	79
Virgo	32	26	14	21	21	23	69	73
Libra	20	22	23	28	24	23	69	76
Scorpio	22	26	25	24	33	26	80	76
Sagittarius	25	21	19	26	38	31	79	74
Capricorn	26	26	28	28	33	29	85	80
Aquarius	29	30	19	28	31	31	79	90
Pisces	45	38	28	27	27	29	102	98
Total	330	335	316	329	331	327	977	991

* Taking the Sun in Aries as an example, the expected frequencies have been calculated as follows:

- Total comparisons between parent and child: 3,923.
- Number of parents (father or mother) born with the Sun in Aries: 358.
- Number of children born with the Sun in Aries: 383.

Therefore the expected frequency for the Sun in Aries (parent **and** child) is 358 × 383 / 3,923 = **35**.

Heredity in Aspects

It is not sufficient to study only zodiacal heredity. We must also take account of aspects, that is, the angular separation along the ecliptic between astrological factors.

For this study the 360° of the zodiac were divided into 36 sectors of 10° each. The first sector, called 0° sector, is centered on the conjunction; the second sector, called 10°, reaches from +5° to 14°59′; and so on. The angular separation in sectors, between parent and child, was then determined for the Sun, Moon, and MC, the expected frequency per sector being 3923 / 36 = 109.

The results are given in Table 3, and show no trace of aspectual heredity. A statistical analysis shows that each distribution is effectively random (chi-square not significant). There is no angular distance that seems more frequent than the others.

TABLE 3

Heredity in Aspects

Number of aspects (Sun-Sun, Moon-Moon or MC-MC) between parent and child. Based on 3,923 comparisons for each astrological factor.

Aspect in Degrees	Sun	Moon	MC	Total
0	128	111	100	339
10	106	108	104	318
20	99	98	121	318
30	100	105	121	326
40	111	104	94	309
50	101	119	114	334
60	123	109	78	310
70	118	104	113	335
80	106	119	120	345
90	106	99	104	309
100	106	115	109	330
110	114	118	108	340
120	138	114	83	335
130	121	101	109	331
140	118	111	110	339
150	97	88	107	292
160	108	118	120	346
170	106	130	121	357
180	101	107	134	342
190	106	102	95	303
200	107	115	114	336
210	92	111	112	315
220	115	113	117	345
230	104	112	104	320
240	98	121	127	346
250	112	103	112	327
260	111	90	109	310
270	99	106	106	311
280	106	133	96	335
290	111	104	102	317
300	85	100	104	289
310	122	106	125	353
320	125	107	97	329
330	113	113	115	341
340	99	102	116	317
350	111	107	102	320
N	3,923	3,923	3,923	11,769
Average N / 36	109	109	109	327

Let us especially consider the results of the conjunction (sector 0°). Paul Choisnard claimed that in 311 comparisons between close relatives (father or mother versus son or daughter) he had found three times as many Moon conjunctions (orb ±5°) as between unrelated persons; the same for the MC; and two times for the Sun. Very striking indeed! But the results of Table 3, for a sample more than ten times larger, reveal no such tendency. In sector 0° there are 111 conjunctions for the Moon, virtually no different from the 109 expected by chance; 100 for the MC; and 128 for the Sun.

For all three factors combined there are 339 cases instead of 327, only marginally in the right direction. None of these differences between actual and average values are significant at the .05 level. The same picture emerges if we look at the other classical aspects in astrology, namely, opposition (180°), squares (90° and 270°), trines (120° and 240°), and sextiles (60° and 300°). It seems that Choisnard's results reflect only random fluctuations due to his small sample size.

Since Kepler, some astrologers also allege that a particular aspect in the parent's birth chart has a tendency to appear in the child's birth chart also. For instance, a parent born with a Sun-Moon conjunction would tend to have children born with the same aspect. I tested this hypothesis using important aspects like Sun-Moon conjunctions, oppositions, and so forth. But I found no significant results. I am forced to conclude that astrologers generally do not realize how much data is needed for testing astrological ideas. For instance, even with 3,923 parent-child pairs, the ±10° Sun-Moon conjunction can appear in both birth charts on the average only 12 times. How can an astrologer be so sure that such heredity connections exist?

Heredity in Houses

In the same way as they divide the zodiacal circle into 12 signs, astrologers divide the diurnal circle into 12 houses. Heredity in houses deserves careful attention because analysis of diurnal motion has already given positive results. As before, I used a large part of the data already gathered for my investigation on planetary heredity (see Appendix 1).

The houses experiment used 16,594 births, that is 8,297 parent-child pairs of birth charts, and involved more than 150,000 planets-in-house computations. The results are shown as observed and expected frequencies in Table 4. In agreement with the findings of planetary heredity, the observed frequency of the Moon, Venus, Mars, Jupiter and Saturn in houses 12 and 9 (which roughly correspond to the rise-culmination zones) is higher than expected, almost without exception.

But apart from this, statistical analysis reveals no evidence for any heredity effect in houses. The differences between expected and actual frequencies do not exceed chance expectation; and the totals in the last column show no significant difference for any case; five of the nine differences are in the wrong direction. Even the grand total gives no significant result, being exactly the same as the expected grand total. Finally we must look at angular houses (I, IV, VII, X), where planets are said to be especially "strong or dominant." But the results of Table 4 (bottom totals) show that children display no tendency to have the same planets in angular houses as their parents, the overall total (2023) being less than expected (2034).

In summary, I could find no evidence in favor of a heredity effect for signs, aspects or houses. Because planetary heredity seems to follow rules that differ from orthodox astrological rules, there is no conflict between these findings and those of planetary heredity.

TABLE 4

Heredity in Astrological Houses*

Number of cases where parent and child have the celestial body in the same house. Based on 8,297 comparisons for each celestial body.

HOUSES →		1	2	3	4	5	6	7	8	9	10	11	12	Total
Sun	Obs	73	73	75	57	46	51	40	51	43	50	71	69	699
	Exp	73	73	75	45	60	41	42	47	47	61	72	70	706
Moon	Obs	70	49	56	52	45	61	46	50	67	51	51	68	666
	Exp	61	57	57	54	57	63	55	60	58	56	55	56	689
Mercury	Obs	75	73	84	57	48	53	45	42	41	65	73	73	729
	Exp	69	79	75	53	48	44	43	42	43	59	71	71	697
Venus	Obs	65	84	64	49	45	38	33	39	65	50	90	91	713
	Exp	72	63	61	57	53	46	39	49	47	59	73	77	696
Mars	Obs	58	61	46	44	52	50	62	59	59	53	63	78	685
	Exp	59	67	51	48	55	51	54	61	60	51	65	69	691
Jupiter	Obs	71	53	63	79	65	66	45	46	63	51	54	57	713
	Exp	59	56	59	62	59	73	50	54	56	53	56	54	692
Saturn	Obs	69	62	61	70	61	78	48	57	59	43	44	54	706
	Exp	71	68	72	70	67	75	47	49	50	42	41	47	699
Uranus	Obs	67	61	70	65	62	72	53	47	48	46	43	37	671
	Exp	71	69	67	66	64	64	51	45	51	48	47	49	692
Neptune	Obs	26	36	36	36	33	34	81	82	98	78	80	80	700
	Exp	34	39	34	37	35	37	82	84	88	81	82	87	720
Total	Obs	574	552	555	509	457	503	453	473	543	487	569	607	6,282
	Exp	569	571	551	492	498	494	463	491	500	510	563	580	6,282

* Taking Mercury in the first house, as an example, the expected frequencies have been calculated as follows:

- Total comparisons between parent and child: 8,297.
- Number of parents (father or mother) born with Mercury in first house: 792.
- Number of children born with Mercury in first house: 727.

Therefore the expected frequency for Mercury in first house (parent **and** child) is 792 × 727 / 8,297 = **69**.

CHAPTER NINE

IS THERE A PHYSICAL EXPLANATION?

First, for the convenience of the reader, let me summarize the main findings. My investigations of over 25,000 deliveries have found that children tend to be born when the same significant celestial body (the Moon, Venus, Mars, Jupiter or Saturn) is in the same diurnal zone (rise-culmination or outside) as it was at the birth of their parents. There is no effect for the Sun or the other planets, or for astrological signs, aspects and houses. The effect is not influenced by sex, duration of labor or birth order. It is increased if both parents share the same diurnal positions. It tends to disappear if the birth is artificially induced. The effect is not large but is reproducible and statistically highly significant.

In Search of an Explanation

Planetary heredity must be examined in the light of scientific discoveries in astrophysics and biology. For example:

— The cosmic environment (cosmic rays, solar particles, magnetic variation, and so on) penetrates into laboratories. It modifies conditions that hitherto have been presumed to be constant.
— This cosmic environment can affect living phenomena. Living organisms are not closed systems but are open to external influences.
— Biological systems are largely colloids in aqueous medium, which are very sensitive to changes in the environment. Therefore even very weak forces are, in principle, sufficient to cause important biological reactions.

— During the birth process, the mother and fetus are in a state of enhanced sensitivity to external influences, which in principle includes cosmic influences.

We must not, of course, assume that what can exist in principle must also exist in fact. Therefore we must accept that the mechanism of planetary heredity remains completely unknown. But as I show below, it is possible to explore various speculations.

Geomagnetism and Planetary Heredity

The challenging question is: could the moon and planets have such subtle influence on humans? The biological effect of the sun is, of course, enormous, yet there is no heredity effect similar to that observed for the moon and the planets. It seems unlikely that the sun plays no role at all; it may have a more general, perhaps indirect, influence; for example, planetary heredity could be the result of disturbances in the solar field caused by the moon and planets, which in turn has an effect on Earth. Such a view is supported by the claims made in 1963 by Australian astrophysicist E. K. Bigg (and others since) that the moon and some planets — Venus and Mars for instance — can influence the geomagnetic field.[1]

These claims are especially interesting because American biologist F. A. Brown has already demonstrated that living organisms react to small variations in the geomagnetic field.[2] So do geomagnetic disturbances, generally related to solar activity, influence planetary heredity? I decided to find out.

For more than a century geomagnetic activity has been recorded every day around the world, and is today expressed in various indices. The earliest index was the **International Magnetic Character Figure**, or **Ci**, which ranges from 0.0 (quiet days) to 2.0 (highly disturbed days, magnetic storms) in steps of 0.1. Ci reflects not field intensity but the degree of disturbance.

For each of the births already gathered for the heredity study that occurred after January 1, 1884, the Ci index was noted. For each of the five significant celestial bodies, the births were then sorted into two categories, namely those showing a heredity effect and those showing no effect. If geomagnetic activity is related to planetary heredity, the

1. E. K. Bigg: "Lunar and Planetary Influences on Geomagnetic Disturbances," *Journ. Geophys. Res.*, 68, 1963, p. 4099.
2. F.A. Brown: "How Animals Respond to Magnetism," *Discovery*, November 1963.

Ci values in the two categories should be significantly different. The results (Figure 16 and Appendix 4) show that is the case. Figure 16 shows that when a planetary effect is present, it increases with increasing Ci; when it is absent, it decreases with increasing Ci.

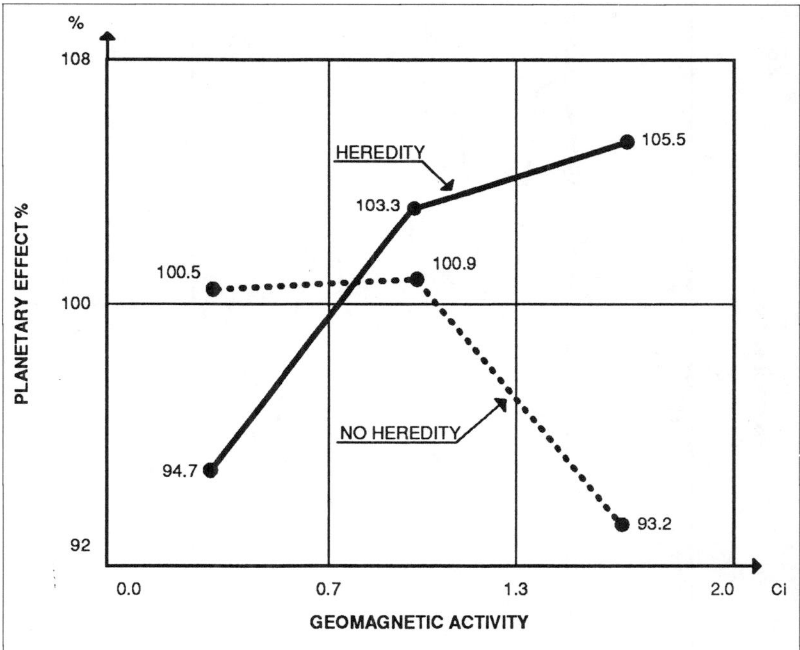

Figure 16: Relationship between Geomagnetic Activity and Planetary Heredity

Horizontal axis: geomagnetic activity Ci divided in three classes.

Vertical axis: intensity of planetary effect in index 100 (100 = average intensity of the effect).

Solid line: the number of (++) cases, i.e., for heredity, increases when Ci increases.

Dotted line: the number of (+ –) cases, i.e., no heredity, decreases when Ci increases.

Combined results for the moon, Venus, Mars, Jupiter and Saturn

The chi-square value for the difference in Ci between the two categories is significant (see Appendix 4). The diurnal patterns are shown in Figure 17. The results show that the planetary effect is roughly twice as strong when the child was born on disturbed days ($Ci \geqslant 1.0$) as when the child was born on quiet days ($Ci < 1.0$).

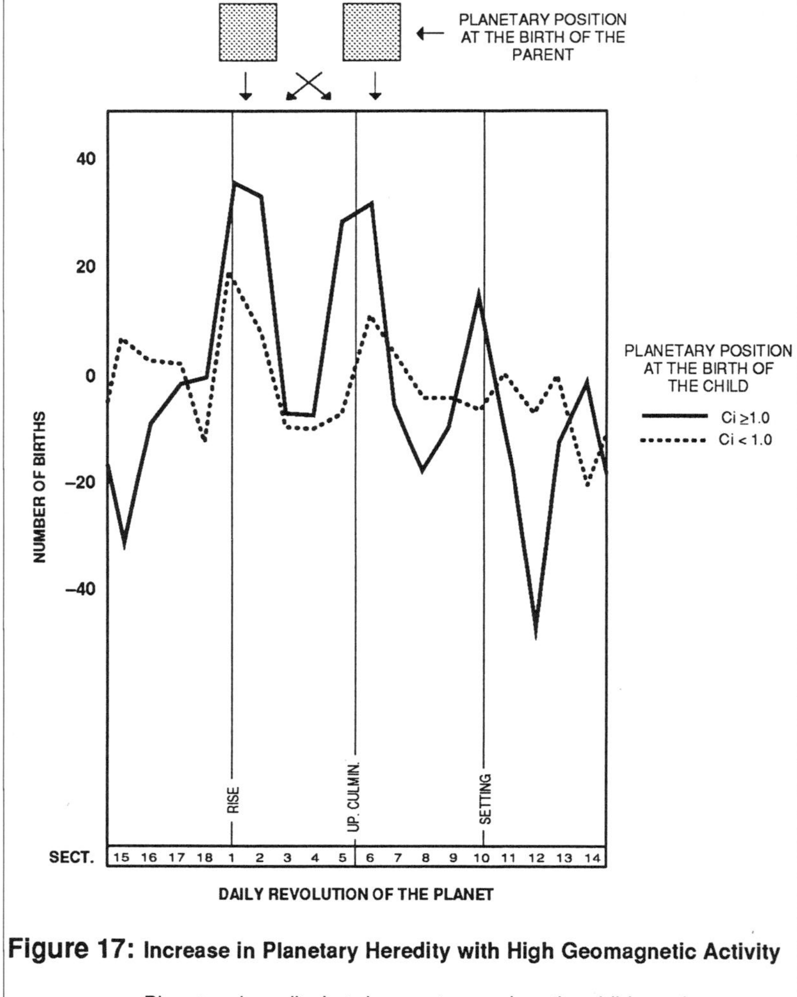

Figure 17: Increase in Planetary Heredity with High Geomagnetic Activity

Planetary heredity is twice as strong when the child was born on disturbed day (Ci≥1.0) as when the child was born on quiet day (Ci<1.0). Combined results for the moon, Venus, Mars, Jupiter and Saturn.

These findings suggest that planetary heredity depends on the solar field, for it is known that geomagnetic disturbances are related to solar activity. It would be interesting to repeat this study with other geomagnetic indices to see if field **intensity** is more important than the degree of disturbance. In the meantime, although the real nature of the effect and its biological mechanism remains unknown, the present results do at least suggest that planetary heredity has a physical basis. (A replication and further studies are described in the Postscript.)

In his book, *Le Signal du Sourcier* (*The Dowser's Signal*), 1962, Prof Yves Rocard, a physicist from the Sorbonne University, mentions the value of magnetic disturbance needed to release the muscular tonus of the dowser so that his rod drops: 50 gammas. Prof Rocard informed me that this value corresponds roughly to the geomagnetic variation when Ci = 1.0, or precisely the value at which the planetary effect in heredity increases. Because the reality of dowsing is disputed, this correspondence may not mean anything, but it is interesting nonetheless.

Cosmic Anthropology?

Planetary heredity helps to eliminate what is perhaps the most absurd aspect of my results with successful professionals, namely the apparent astral determinism at birth. With planetary heredity there is no such determinism; only genetic factors are determinant, and the planet is no more than a minor participant in the birth process. As a result the planet becomes merely an unwitting (and rather unreliable) indicator of our genetic makeup. Planetary heredity also shows that planetary effects apply to ordinary people as well as to famous people. With the discovery of planetary heredity, my previous observations have lost a great deal of their occult perfume. But people will still ask: "Why a planetary heredity? Do we need it to be what we are?" To this question my only answer for the moment is: "The universe is fantastic, don't you realize that?"

Michel Gauquelin
Paris, 1960-1965

POSTSCRIPT

THE 1988 STATUS OF PLANETARY HEREDITY

Since the original publication of this book in 1966, I have worked hard for a better understanding of planetary heredity. In this Postscript, I would like to briefly summarize the fruits of my efforts. It is divided into four sections:

1. Exploring new directions of research (1966-1970)
2. Replicating planetary heredity (1972-1977)
3. Computer re-analysis of the data (1979-1984)
4. The puzzling 1984 planetary heredity experiment

1 — Exploring New Directions of Research (1966-1970)

GEOMAGNETISM AT THE BIRTH OF PARENTS

I found previously that the heredity effect on the **child** is roughly twice as strong on geomagnetically disturbed days ($Ci \geq 1.0$) as on quiet days ($Ci < 1.0$). What about geomagnetic activity at the birth of the parent? To find out, I took the same data and noted the Ci index at the birth of each parent. All details about the procedure can be found in Postscript reference 1.

The results (Figure 18) are intriguing, even puzzling. As Ci increases at the birth of the child, the heredity effect increases more or less smoothly. But as Ci increases at the birth of the parent, the heredity effect merely fluctuates and then decreases sharply. The difference is particularly marked at high values of Ci, although this may, of course, merely reflect random fluctuations due to the small sample sizes at high Ci. Examination of the data (Ci parent/Ci child using a 2 × 2 table) showed that cases of planetary heredity were least frequent with Ci ⩾ 1.0 (parent) and Ci < 1.0 (child). If real, this observation might be interesting. It goes without saying that these experiments should be replicated on a larger sample and, if possible, with the help of a computer. (I should perhaps remind the reader that I carried out all these experiments by hand.)

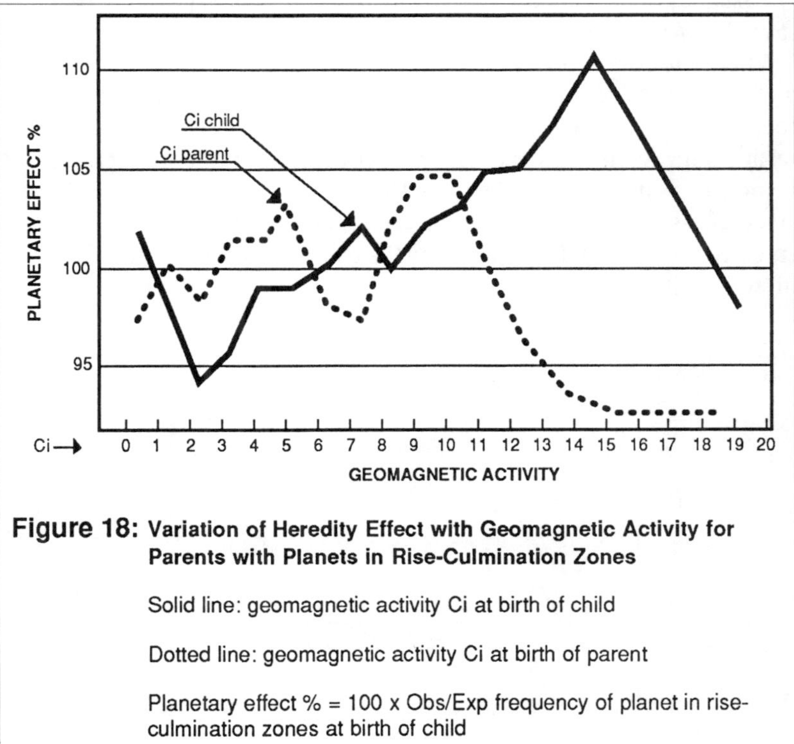

Figure 18: Variation of Heredity Effect with Geomagnetic Activity for Parents with Planets in Rise-Culmination Zones

Solid line: geomagnetic activity Ci at birth of child

Dotted line: geomagnetic activity Ci at birth of parent

Planetary effect % = 100 x Obs/Exp frequency of planet in rise-culmination zones at birth of child

SUNSPOT NUMBER AND PLANETARY HEREDITY

Because geomagnetic activity is correlated with solar activity, planetary heredity may be under the control of solar activity. To test this idea, I used the classical measure of solar activity, namely, the sunspot relative number **R**, which has been recorded daily by astronomers since 1818 (cf. M. Waldmeir, *The Sunspot Activity*, Zürich, 1961). For each of the births of children already gathered, the value of **R** was noted. I then sorted the **R** values into intervals of 10 unit each from **R** = 0 (no sunspots) to **R** = 130 and above (many sunspots).

Methodology, and actual and expected frequencies, can be found in Postscript reference 2. The results (Figure 19) showed an increase in planetary heredity for sunspot numbers above 90 or so, that is, when solar activity is high. The effect is similar to that observed with Ci but is more erratic (Figure 19) probably because **R** is a reliable long-term measure but an unreliable short-term measure; that is, it does not accurately reflect the influence of solar activity on the Earth at any given moment. Thus the correlation between **R** and Ci is very high on a yearly basis but not on a daily basis. American astrophysicist S. Chapman says (in *Magnetism and the Cosmos*, Oliver & Boyd, London, 1967): "When one compares daily values of R and Ci one observes that the difference between them can be marked." Astrophysically the reason for this is easy to explain (see Postscript reference 2). At any rate, my results suggest the existence of solar control on planetary heredity, **R** being a less satisfactory index than Ci for showing it. But again we must be cautious about interpreting such observations. They need to be replicated first on a larger sample of births using, if possible, a better index of solar activity.

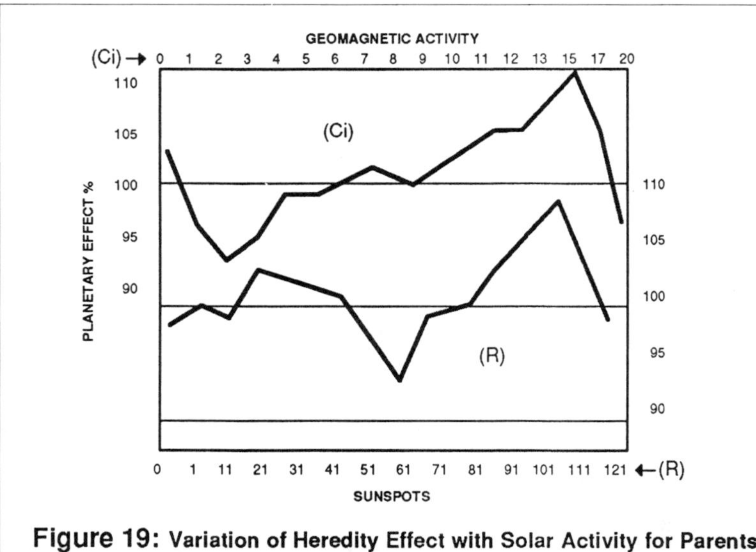

Figure 19: Variation of Heredity Effect with Solar Activity for Parents with Planets in Rise-Culmination Zones

Upper curve: geomagnetic activity Ci at birth of child (from Figure 18)

Lower curve: daily sunspot number **R** at birth of child

Planetary effect % = 100 x Obs/Exp frequency of planet in rise-culmination zones at birth of child

PLANETARY HEREDITY AND THE DISTANCE EFFECT

I suggested previously that the planetary effect may vary according to the visibility or distance of the celestial body. So is it possible to observe a variation in planetary effect as orbital motion varies the planet-Earth distance? This distance variation may be small, as for the moon, or it may be very large, as for Venus and Mars. Astronomers measure planet-Earth distance in astronomical units (a.u.), 1 a.u. being the mean distance between the Earth and sun. The distance from Earth is 0.3-1.7 a.u. for Venus and 0.4-2.7 a.u. for Mars, so it can vary by a factor of 6 or 7 in both cases.

If planetary effects vary as, say, the square of the distance, then Venus and Mars are good candidates for demonstrating it. Using information taken from the French *Annuaire du Bureau des Longitudes*, Paris, I noted the Venus-Earth and Mars-Earth distances at the birth of each child in my heredity sample. I then sorted the distances into intervals of 0.1 a.u. and looked at the heredity effect for each interval. Methodology, and actual and expected frequencies, can be found in Postscript reference 3.

The results (Figure 20) showed no clear distance effect for either Venus or Mars, the trends being, if anything, in the wrong direction. So, contrary to my expectation, planetary heredity does not vary according to planetary distance. This result is not only puzzling in itself, but it is also in disagreement with the previous finding that only the five celestial bodies closest to the Earth display an effect (Chapter Four.).

SUN-PLANET ANGULAR SEPARATIONS

Does the Sun-planet angular separation at birth affect planetary heredity? To find out, I sorted the Sun-Venus and Sun-Mars angular separations in my heredity sample into intervals of 5° for Venus and 20° for Mars. Methodology, and actual and expected frequencies, can be found in Postscript reference 3.

The results (Figure 21) are not significant but they present some interesting features that should be tested again with more data. For Venus, the planetary effect seems to be weakest when the planet is at maximum elongation, which is rather suggestive although the difference between actual and expected frequencies does not reach significance. For Mars, there is a zigzag pattern that tends to show minima around Sun-Mars conjunction (0°-20°), Sun-Mars square (80°-100°), and Sun-Mars opposition (160°-180°). But as for Venus, these observations are more suggestive than conclusive. This study deserves to be completed by the analysis of Sun-Moon, Sun-Jupiter and Sun-Saturn

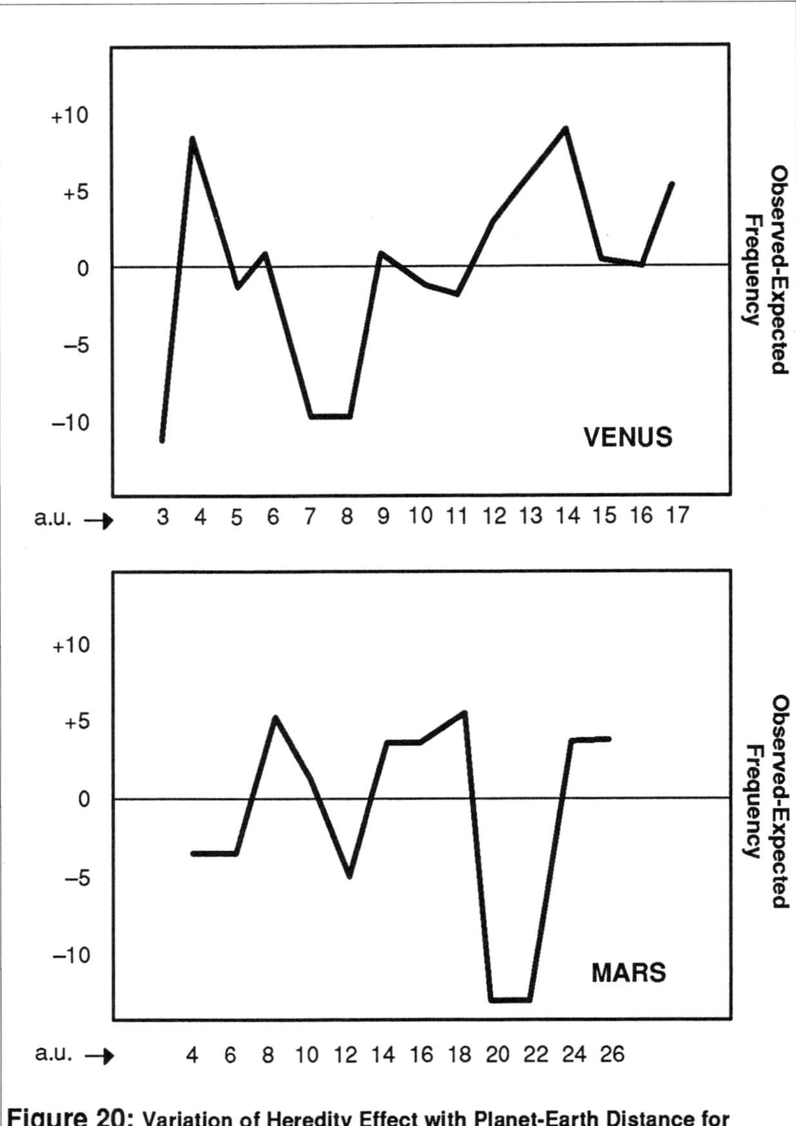

Figure 20: Variation of Heredity Effect with Planet-Earth Distance for Parents with Planets in Rise-Culmination Zones

Upper graph, Venus. Lower graph, Mars.

Horizontal axis: planet-earth distance in a.u. at birth of child.

Vertical axis: observed-expected frequency of planet in rise/ culmination zones at birth of child.

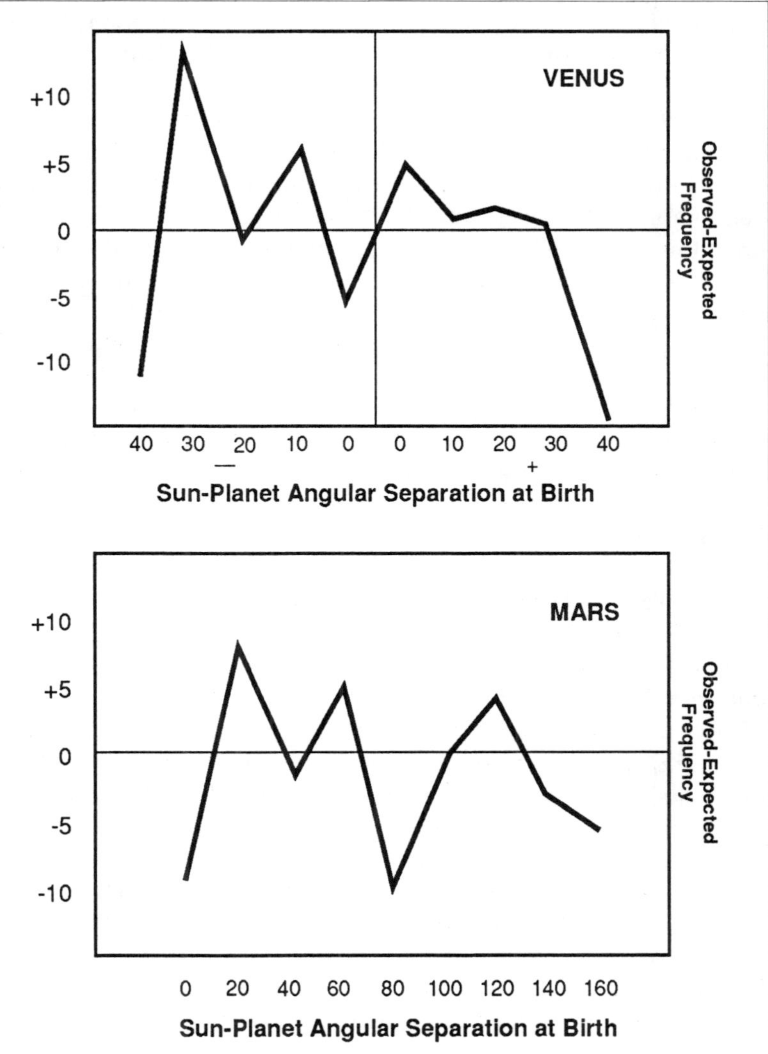

Figure 21: Variation of Heredity Effect with Sun-Planet Angular Separation for Parents with Planets in Rise-Culmination Zones

Upper graph, Venus. Lower graph, Mars.

Horizontal axis: sun-planet angular separation in degrees at birth of child. For Venus this value cannot exceed 48° for astronomical reasons.

Vertical axis: observed-expected frequency of planet in rise/culmination zones at birth of child.

angular separations, but in 1968 this was impossible because it was extremely time-consuming to carry out the endless calculations by hand. Hopefully this work will now be undertaken by computer.

SETTING AND LOWER CULMINATION ZONES

I found previously (Figure 8) that the setting and lower culmination zones seem to weakly echo the effects found for rise and upper culmination zones. But no statistical analysis of the data was done at that time; this appeared later in 1968 and can be found in Postscript reference 4. This analysis showed that the effect in setting and lower culmination zones, although weak, was significantly different from that in sectors outside the rise and upper culmination zones. For example, children with parents born with Mars in rising or upper culmination zones tend to be born with Mars in the same zones (difference between observed and expected frequencies +85) or in setting or lower culmination zones (difference a weak +2), and tend **not** to be born in sectors in between (difference −87). The effect for the Moon, Venus, Jupiter and Saturn is more or less the same.

PARENTS WITH ONLY ONE PLANET IN RISE-CULMINATION ZONES

In 1969 I looked at what happens when the parent was born with **only one** significant planet in rise-culmination zones (N = 6,346) instead of two or more. Because the planet has no interference from others, as it were, one might expect a stronger planetary heredity for the child. Methodology and computations can be found in Postscript reference 5.

The results did not support the expectation. The heredity effect showed no significant dependence on the number of planets in rise-culmination at the birth of the parent. Although the result is not significant, its interpretation would be premature. But we may note that a complementary study is possible, namely the analysis of cases where the child (and not the parent this time) was born with only one planet in rise-culmination zones. Such a study deserves to be carried out.

2 — Replicating Planetary Heredity (1972-1977)

Several years after the publication of *L'Hérédité Planétaire*, I decided to replicate the heredity experiments involving the times of birth and the onset of labor.

These new experiments were performed in the city of Bourges for the birth of children in the years 1923-1939, and in the 14th arrondissement of Paris for the birth of children in the years 1923, 1924, 1929 and 1930 (the 14th arrondissement has the largest concentration of maternity hospitals in Paris). In Bourges it was unfortunately not possible to get access to the medical records, so only a replication of planetary heredity at birth has been done. In Paris, thanks to the help of Prof Lepage and Prof Sureau, medical records were made available at the largest maternity hospital (Baudelocque) in the 14th arrondissement; so I was able to directly compare the birth records from registry offices with the medical records giving time of onset of labor, whether induced, weight of baby, age of mother, number of previous births, and so on. The data was gathered almost entirely by collaborators working under my direction: in Bourges, at first by Jacques Reverchon then by the staff of the registry office; in Paris by Huguette Lampe, Annie Pillard and Michele Paruta. Planetary computations were made by hand by Jacques and Aimee Reverchon. Altogether data on 25,694 births were gathered (35% from Bourges), that is 18,556 parent-child comparisons, or 16% more than before. The results were published in 1977 by my laboratory (see Postscript reference 6). But the results concerning the onset of labor (N = 9,643 or 154% more than before) remained unpublished until now. In both cases the methodological procedures were exactly the same as for the original experiments described earlier.

TIME OF BIRTH

The results of the new experiment, like those of the previous one, show a positive planetary heredity effect for the Moon, Venus, Mars, Jupiter and Saturn (Figure 22), and not for the Sun and other planets. However, there are some differences. For example, the effect for Jupiter is stronger than before (Figure 7), whereas the effect for Venus and Saturn is weaker. However, the diurnal patterns remain very similar, so these differences may be due to random fluctuations.

The other main characteristics of planetary heredity are also supported by the new results. The effect is not influenced by the sex of parent or child; it increases if both parents have the same planetary heredity; and it disappears if the birth is artificially induced. The effect seems again to be related to geomagnetic activity, but less clearly than before, so on this point new data are needed. Overall then, the replication is successful. So far, so good.

TIME OF ONSET OF LABOR

In general, the results of the new experiment are negative. Despite a much larger sample, only Venus shows a (marginally) significant result, whereas the results for Jupiter and Saturn are slightly against the hypothesis. The chi-square value for the five planets combined is not

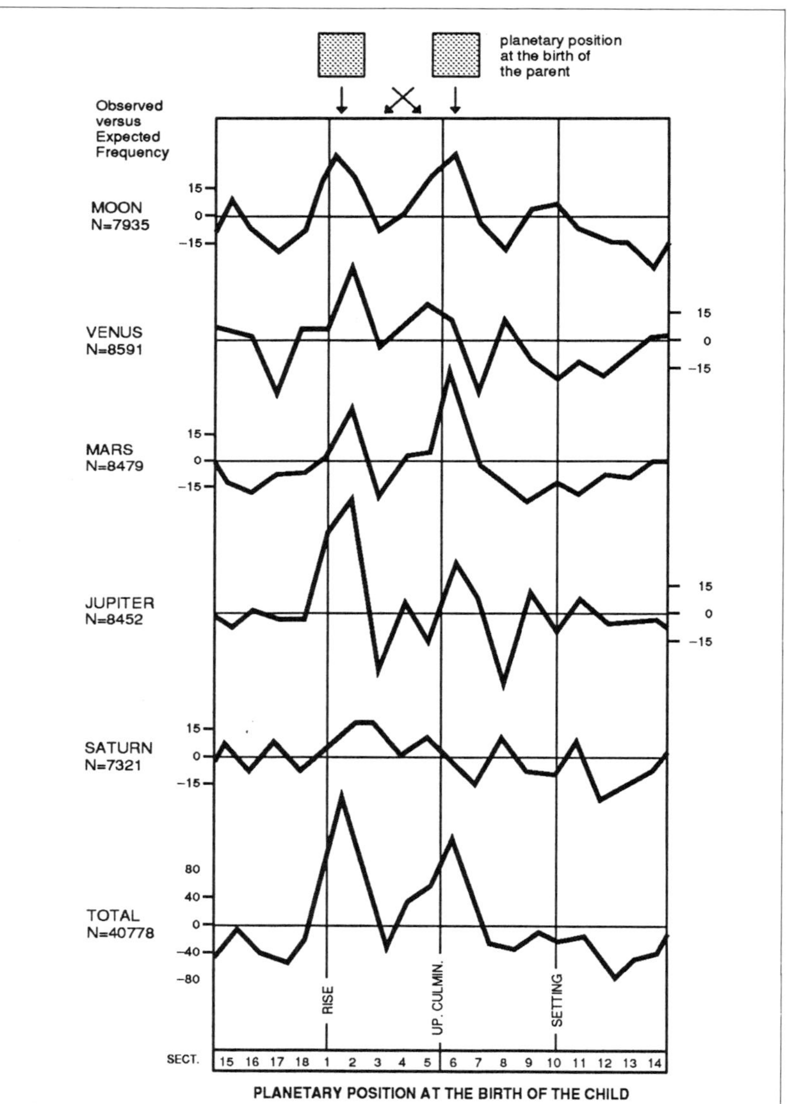

Figure 22: The 1976 Replication
Planetary Heredity for the Moon and Planets Combined

Horizontal axis: planetary position in 18 sectors at birth of children when the parents have the same planet in one of the two plus zones indicated by the shaded boxes. **Vertical axis**: difference between observed number of children per sector and expected number. (from Gauquelin, reference 6, 1977)

significant. In other words, for this new sample, for parents born with a significant planet in rise-culmination zones, labor does not tend to start when the same planet is in the same position. The diurnal pattern (Figure 23) is erratic, contrary to that previously observed (Figure 14), and shows no excess of onset of labor in rising or culmination zones.

It is possible that the accuracy of times of onset in the rather old Baudelocque records (1923-1929) can be questioned. Nevertheless, I think that the results of this new experiment, which required much time, energy and money, cannot reasonably be considered as encouraging for the hypothesis under investigation. There is no support for the theoretical model I suggested to explain the biological mechanism of planetary heredity (see Chapter Seven).

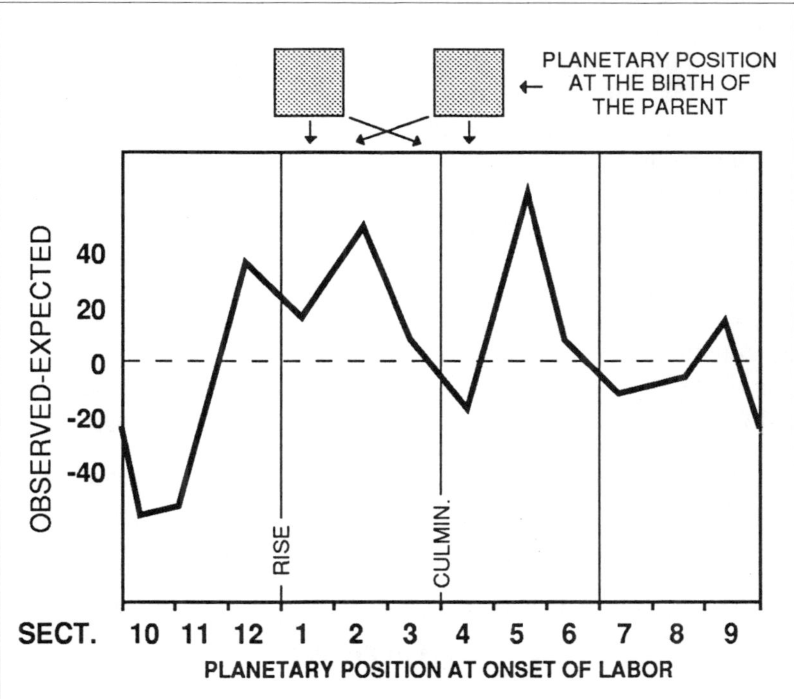

Figure 23: Planetary Heredity and Onset of Labor (Second Experiment)

The previous results (Figure 14) do not seem to be replicated by this second experiement.

Graph shows difference between observed and expected frequencies for the five significant planets combined.

3 - Computer Re-analysis of the Data (1979-1984)

During the 1979-80 winter in San Diego, with the help of Astro Computing Services, I carried out a re-analysis by computer of all the birth data previously calculated by hand in the 1966 and 1976 heredity experiments. Full details can be found in Postscript references 7 and 8.

For both experiments the computer results confirm my original results but are somewhat lower in significance, probably because of inadvertent bias in the hand calculations when interpolating from tables of sectors (see Appendix 2). The loss of significance is much greater in the second experiment, and is marked for all five planets, the significance for all planets combined falling from .0001 (hand calculation) to .01 (computer re-analysis). This is particularly disturbing because the hand calculations were done by the Reverchons, not by myself, in order to obtain results free from bias. I even requested the Reverchons to perform all the calculations twice (and paid them in consequence), my role being limited to checking any discrepancies.

As well as checking the hand calculations, the computer confirmed all the previous findings and made some new ones. The area showing the maximum planetary effect was confirmed to be the rise-culmination zones shown shaded in Figure 5; elsewhere I refer to these shaded zones as "plus zones." If a parent is born with a planet in rise-culmination zones, there is no significant tendency for the child to be born with a **different** planet in the same zones or outside them; this suggests that each planet operates independently of the others.

But I was worried by the re-analysis of the Reverchons' hand calculations. So in 1981 I decided to undertake a third experiment on a very large scale, this time with the computer doing all the work.

4 - The Puzzling 1984 Planetary Heredity Experiment

It took me and my collaborators — Genevieve Artru, Maryse Damiens, Valerie Loizance and Genevieve Martichoux — three years of hard work to gather 50,942 birth data from five separate areas, four from Paris and one from Lille in the north of France. It includes 33,120 parent-child comparisons, almost as many as in the two previous experiments combined. All planetary computations and the statistical analyses were performed in San Diego by the computer of Astro Computing Services. Full details can be found in Postscript reference 8.

Sadly, the results of this huge new experiment showed only slight agreement with those of the previous two experiments (Figure 24). This time, except perhaps at the rising zone, there is almost no tendency for parents born with a significant planet in rise-culmination zones to have children with the same planet in the same zones.

The distribution of the birth times is normal and shows no evidence of medical interference, so if planetary heredity is real, then the data should show it. But the chi-square values for each individual planet, and for all five planets combined, are not significant.

The results for both parents born with the same planet in rise-culmination zones are a little more encouraging, but the effect remains non-significant and is much weaker than before. The results for geomagnetic activity are also encouraging, and tend to support the hypothesis, but again they are not significant.

There is no doubt that these observations constitute a setback that is hard to reconcile with the significant and consistently positive results that I observed over 25 years of work, especially in view of a recent comparison of planetary heredity between two samples from the same (12th) arrondissement of Paris. The samples were for 1923-31 and 1931-39 respectively; each consisted of nearly 12,000 birth comparisons, and each had been checked by the same people. The samples seemed identical in every way, yet the first gave results strongly in support of the hypothesis while the second gave results almost as strongly against the hypothesis (Postscript reference 9). Puzzling indeed!

Now, of course, we have to ponder all that was found. Careful cross-examination of past and present materials might explain the disquieting discrepancies. I agree when scientists say that "facts are always right," so we should be able to find an adequate explanation.

The 1988 status of planetary heredity is therefore an uncertain one. In the discussion of my 1984 report I concluded (after the last negative results): "The only scientific attitude would be to replicate the experiment again on 100,000 fresh birth data. As we are working with non-eminent people for this kind of investigation, the human material is almost inexhaustible. For my part, I ought to leave the burden of this huge study on shoulders other than mine. Volunteers are most welcome!" More than three years have passed since then. Quite understandably, nobody has volunteered. But time, in science, seldom flies. It passes very slowly. So we must wait. One day we will know the truth about planetary heredity. In the meantime, let me say that, despite the disappointments, I am pleased about one thing: if my latest work has created doubt, then this is the best that can happen in science. As Bertrand Russell has written, "Not to be absolutely certain is, I think, one of the essential things about rationality." (Postscript reference 10).

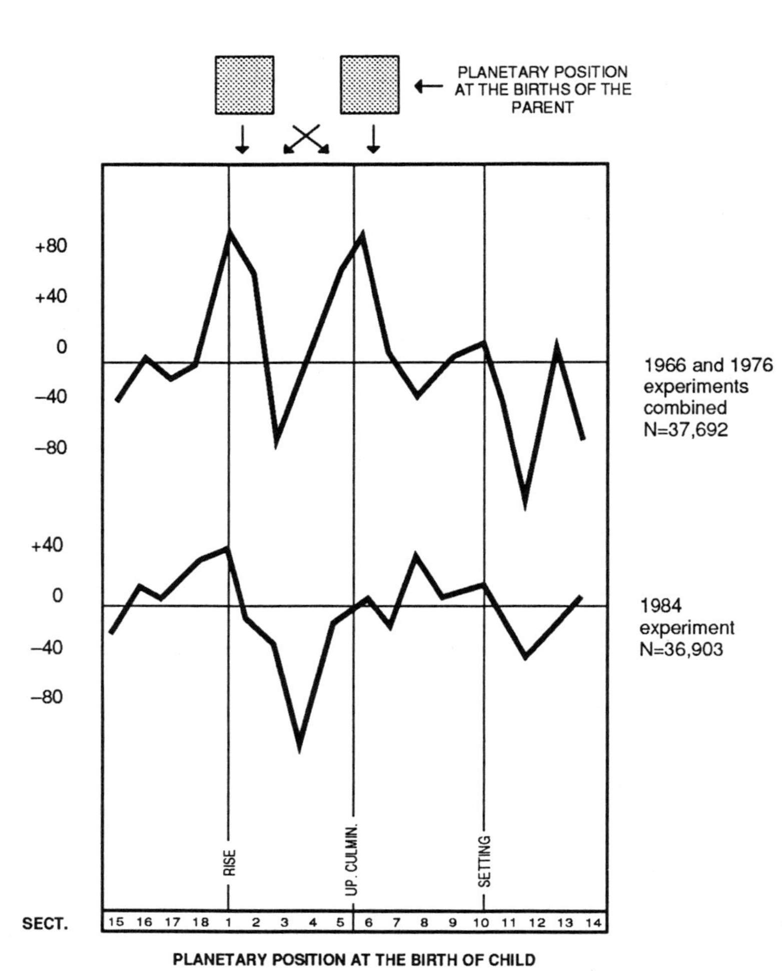

Figure 24: Planetary Heredity for the Moon and Significant Planets Combined

Horizontal axis: planetary position in 18 sectors at birth of children when the parents have the same planet in one of the two plus zones indicated by the shaded boxes.

Vertical axis: difference between observed number of children per sector and expected number.

(from Gauquelin, reference 8, 1984)

References to Postscript

1. M. Gauquelin (1968): "L'effet planétaire d'hérédité et le magnétisme terrestre," *Les Cahiers Astrologiques*, 134:459-473.
2. M. Gauquelin (1969): "Nombre relatif journalier des taches solaires et l'effet planétaire d'hérédité," *Les Cahiers Astrologiques*, 143:259-262.
3. M. Gauquelin (1968): "L'effet planétaire d'hérédité en fonction de la distance de Vénus et de Mars à la Terre," *Les Cahiers Astrologiques*, 136:561-569.
4. M. Gauquelin (1968): "L'effet planétaire d'hérédité dans les secteurs du coucher et de la culmination inférieure," *Les Cahiers Astrologiques*, 134:451-458.
5. M. Gauquelin (1969): "Un cas particulier dans l'effet planétaire d'hérédité," *Les Cahiers Astrologiques*, 141:136-138.
6. Michel & Françoise Gauquelin (1977): *Replication of the Planetary Effect in Heredity*, Series D. Volume 2, LERRCP, Paris.
7. M. Gauquelin (1984): "Profession and Heredity Experiments: Computer Re-analysis and New Investigation on the Same Material," *Correlation*, 4 (1):8-24.
8. M. Gauquelin (1984): "Planetary Heredity: A Reappraisal on 50,000 Subjects," *New Birthdata Series*, Volume 2, LERRCP, Paris.
9. M. Gauquelin (1985): "On the Lack of Positive Replication of Planetary Heredity," *Correlation*, 5(1):36-42.
10. M. Gauquelin (1986): "Bibliographic Chronology of Michel Gauquelin's Planetary Heredity Research," *Correlation*, 6(1):16-17.

APPENDICES

METHODOLOGY AND RAW RESULTS

In these appendices I describe the methodology and raw results necessary for validation of my statistical calculations by others. The material is exactly as presented in the original 1966 edition, with a few small additions to relate it to later studies, and is broken down as follows:

Appendix 1. The Gathering of Birth Data
Appendix 2. Astronomical Problems
Appendix 3. Statistical Analysis
Appendix 4. Tables of Raw Results

APPENDIX 1

The Gathering of Birth Data

DIFFICULTIES IN GETTING INFORMATION

At first sight, it is not difficult to get birth data from official birth records for testing the planetary hypothesis. In fact, we met great difficulties. In France the time, date and place of birth are recorded on the birth certificate and are available to anyone who asks for it (and pays the fee). But other details, such as names of parents, are not available for reasons of privacy. Our efforts, therefore, had first to be directed toward securing official authorizations to consult the records. Our requests were not always successful, which partly explains the various geographical origins of our data.

Another difficulty arose because birth records before 1923 do not provide information concerning the parents. For instance, in Paris and its suburbs the birth records included the parents' name and age, but not their date and place of birth. Fortunately, since January 1, 1923, a law has made it obligatory to provide this information in all birth records. Thus, for children born since 1923, it became possible to request the parents' birth data by mail from the registry office at their birthplace. That explains why most of our heredity experiments start from January 1, 1923.

FIRST EXPERIMENT IN PARIS

The heredity research started using the birth registers of a large Parisian district, the 12th arrondissement. We gathered 3,600 birth data of children, from January 1, 1923, whenever **one or both of their parents** were born in the same arrondissement. Thus we could find without difficulty (and without spending money) the birth time of the parent from the birth records in the same office.

A second sample of parent-child pairs was then gathered, starting again from January 1, 1923. This time we noted the first 1,200 births where **both parents** were born in places other than Paris, requesting by mail the birth data of the parents from the registry office at their birthplace. We intended thus to gather a sample of 2,400 parent-child pairs. But we obtained complete information for only 2,303 cases, for several reasons: some records gave no time of birth, some records could not be located by the registry office, and some offices refused to provide information (fortunately this was rare). We could not afford to use subjects for whom the registry office required a special research fee, so the sample was limited to data obtainable only for the price of a return stamp. Fortunately this was most often the case. All exclusions are documented, for all letters received from registry offices are filed in alphabetical order in my laboratory. In this way we were able to gather in the 12th arrondissement a total of 5,903 comparisons between parent and child.[1]

Results of this first study have been very encouraging and we tried to replicate the experiment on other samples of births.

EXPERIMENTS IN PARISIAN SUBURBS

Unfortunately, we could not get authorization to work in another Parisian arrondissement. However, owing to the very kind and efficacious mediation of one of my friends, we were able to get permission to consult the birth registers of five Parisian suburbs, namely Nogent-sur-Marne, Créteil, Joinville-le-Pont, Vincennes and Montrouge.

At Créteil, Joinville-le-Pont, Vincennes and Montrouge, we gathered the birth data of all children born between January 1, 1923, and December 31, 1945, whose parents (father, mother or both) were **born in the same locality** as the child, so that (as in the first Parisian experiment) we could look in the same birth records for their birth-time. We did not include the births occurring after 1945, because by then the proportion of induced births has started to increase.

A fortunate situation in Créteil greatly increased our sample. Since the end of 1937, this town had possessed a large hospital with an important maternity ward, so great numbers of mothers from surrounding cities have been delivered there. So in this case we were able to include children whose parents had been born **elsewhere**. We also attempted to gather birth data occurring before 1923, in order to have a sample with more natural deliveries. This was possible in Créteil, Joinville-le-Pont, and Nogent-sur-Marne. Although birth records before 1923 give only age and name of parents, without date and place of birth, the population of these suburbs was rather stable at that time, so a fair number of parents were born in the same place as their children. After much tedious work we were able to locate the birth records for approximately one case out of ten, resulting in 2,220 more parent-child pairs. However, parents born before January 1, 1850, were not included because the birth times in such old records seem insufficiently accurate for the purpose of heredity work. In Nogent-sur-Marne,

1. These data (9,846 birth times) have been published in the two first volumes of our *Series B* (LERRCP, Paris, 1970).

the births were gathered between 1850 and 1900 only, because the registers for more recent births were not available. In this way we were able to gather in the area surrounding Paris an additional 10,134 comparisons between parent and child.[2]

SUMMARY

The data gathered in Paris and its surroundings totaled 16,027 parent-child pairs. In 3,487 cases the child could be compared with both parents, father **and** mother (6,974 comparisons). In 9,063 cases the child could be compared with one parent only, father **or** mother, because information for the other parent could not be obtained.

The data included 3,837 comparisons between father and son, 3,471 comparisons between father and daughter, 4,433 comparisons between mother and son, 4,296 comparisons between mother and daughter, and 6,691 comparisons between siblings.

INFORMATION ON DELIVERIES

For information concerning the delivery conditions, the birth registers cannot be of any help. We need instead to consult the **medical** records. But for old births, which usually occurred at home, medical records have never existed. Even if the births had occurred in a hospital, the records may have been destroyed, as was the case for births before 1931 at the Saint-Antoine hospital in the 12th arrondissement of Paris, where the first experiment on heredity was undertaken.

This is why no medical information concerning the births of parents was found, and why information for only three thousand children was found. Fortunately, the parents were born at a time when the deliveries were generally natural, as the distribution of their births during 24 hours shows. For the children this is not always the case, which is why it was useful to get medical information about the delivery conditions. At Créteil Hospital we were able to examine all the medical records since it was founded at the end of 1937 until December 31, 1945. To the birth data already gathered from the registry offices, the following medical items were added: time of onset of labor and its duration, the mother's age, the number of previous births, the presentation of the fetus, the weight of the baby, and the medical interventions (if any) that happened during the delivery.

---

2. These data (15,115 birth times) have been published in the four last volumes of our *Series B* (LERRCP, Paris, 1970-71).

INFORMATION ON DEATHS

For practical convenience we gathered official death data for subjects who died in the same place as they were born. Three Parisian suburbs were investigated, namely Vincennes, Charenton-le-Pont and Montrouge. As before, we discarded subjects born before January 1, 1850, due to the uncertain accuracy of their birth times. The deaths occurred between January 1, 1900, and December 31, 1955, in Vincennes; and between January 1, 1900, and December 31, 1945, in Charenton-le-Pont and Montrouge.

Altogether a total of 4,779 birth-death comparisons were gathered (2,587 males and 2,192 females). The sample is not representative of the general population because half the subjects (2,244 cases) died before the age of two. The younger the subject the greater the chance of dying in the birthplace, which explains the abnormally high proportion of young deaths. For the same reason it is possible that the proportion of accidental death is also higher than average, which is unfortunate because, according to the hypothesis, the study should test natural death only. Some accidental deaths were easily identified and discarded because they occurred in the battlefield during World Wars I and II; in such cases the birth records contain the words "Mort pout la France" (Dead for France).

APPENDIX 2

Astronomical Problems

Astronomical data were obtained from well-known French astronomical publications like *Annuaire du Bureau des Longitudes, Annuaire de Flammarion*, and *Connaissance des Temps*, and to a lesser extent from ephemerides made for astrologers.

THE DIURNAL MOVEMENT OF THE PLANETS

Each day, as a result of the earth's rotation, all celestial bodies appear to move in a 24-hour trajectory called the diurnal movement.

Consider, for instance, the diurnal movement of Mars as seen from Paris on May 24, 1956. In the *Annuaire du Bureau des Longitudes* we find that on that day, in Paris, Mars rose at 0 hour 44 minutes, culminated at 5 h 33 m, and set at 10 h 22 m, describing the circle ABCD

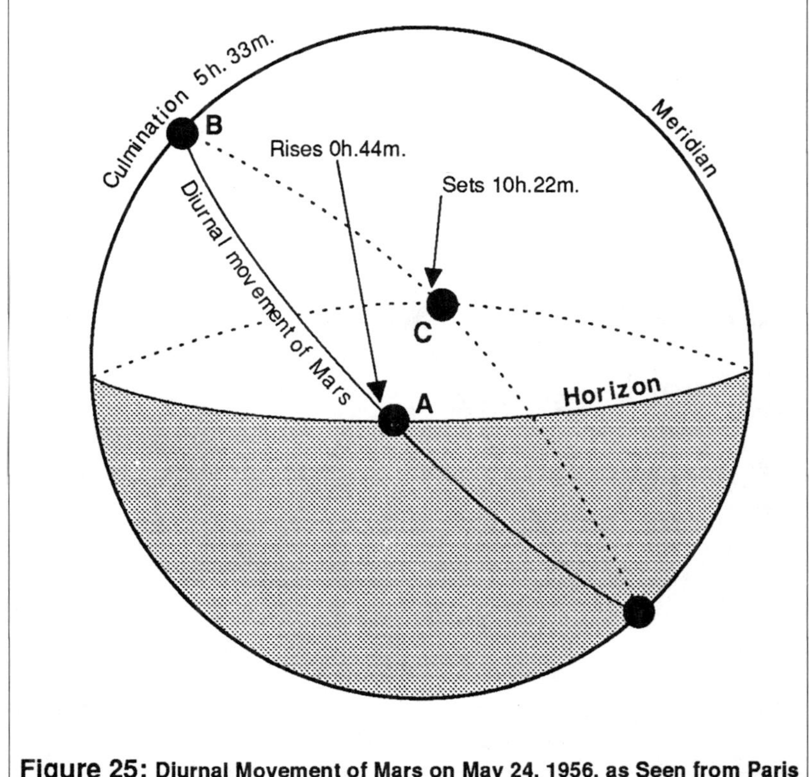

Figure 25: Diurnal Movement of Mars on May 24, 1956, as Seen from Paris

shown in Figure 25. Beneath the earth, Mars reaches its lowest point when it again crosses the meridian at D, subsequently to reappear above the horizon close to A.

Because the Earth's rotation is uniform, so is the diurnal motion of Mars. Thus, AB takes the same time as BC, and CD takes the same time as DA. (But AB and DA are usually different, in the same way that day length and night length are usually different.) Thus the position of Mars, or any other planet including the sun and moon, can be accurately interpolated for any given time.

DIVISION OF THE MOVEMENT INTO SECTORS

Division into sectors was explained in Chapter Two and is further illustrated in Figure 26. Here the diurnal (daytime) arc of Mars is 578 minutes, and its nocturnal (nighttime) arc is 862 minutes. Suppose that we want to divide the whole diurnal movement into twelve sectors. Mars will remain in each diurnal sector for 576 / 6 = 96 minutes, and in each nocturnal sector for 862 / 6 = 144 minutes. The figure shows the times at which the planet passed from one sector to the next.

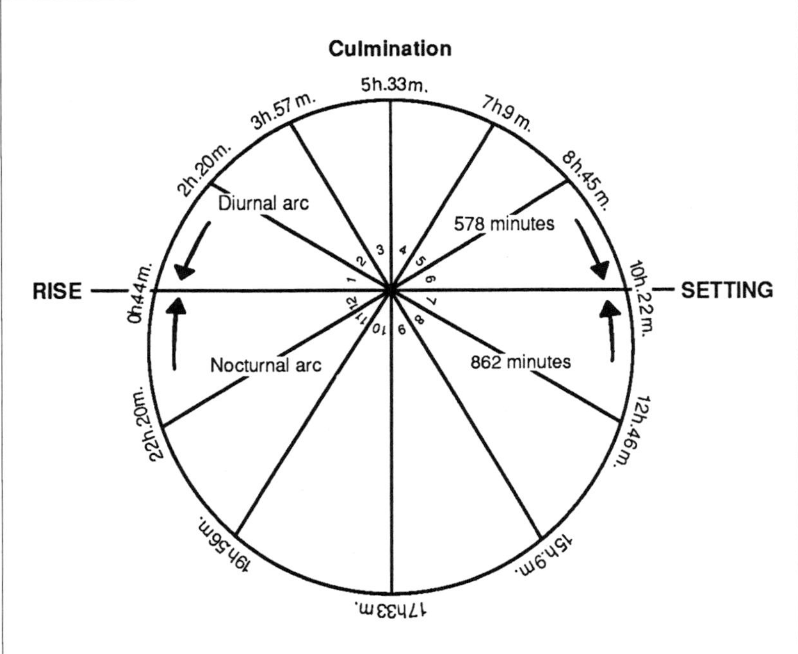

Figure 26: Division of the Diurnal Movement of Mars into Twelve Sectors on May 24, 1956, as Seen from Paris

If a person had been born at 1:00 AM on that day, we would then say that this person was born with Mars rising, or, more accurately, with Mars in sector 1. Similarly, if the birth had occurred at 6:00 AM, we would say that Mars was in sector 4, instead of saying it had just culminated. In a sample of several thousand births, we expect a few hundred with Mars in sector 1, a few hundred with Mars in sector 2, and so on. The same applies to the other celestial bodies. In this way the pattern of births during the diurnal movement can be determined.

In my research, I divided the daily orbit of each celestial body into 36 sectors. This allows regrouping into 18, 12 or 9 sectors for particular analyses. Most of the graphs in this book are based on 18 sectors.

TABLES OF SECTORS

In principle, the above methodology is simple. In practice, it is tedious and requires a great number of calculations. So I gave up and decided, from the beginning of my research, to use tables of sectors calculated in exactly the same way as for Placidus houses in astrology, except there are 36 clockwise sectors instead of 12 counterclockwise houses. With such a table one gets, in a single calculation using the sidereal time and geographical latitude, the sector occupied by the ecliptic position of each celestial body, as illustrated in Figure 4. Interested readers can find more details in Michel and Francoise Gauquelin's book of *Methodes*, Paris, 1957.

Our "Placidus" sector tables give, for every 12 minutes of sidereal time, the 36 diurnal sectors in degrees of ecliptical longitude. Four tables were computed for our purpose for geographical latitudes of 40°, 45°, 49° and 53° North. Two of them — for 45° and 49° — appear in our book of *Methodes* (pp. 98-109). This simplified method ignores zodiacal latitude, but careful comparison with the results using true rise/set sectors reveals no significant discrepancy or bias. (Michel Gauquelin, "La Latitude Zodiacale," *Les Cahiers Astrologiques*, 92, May 1961, pp. 119-129; "Profession and Heredity Experiments: Computer Re-Analysis and New Investigations on the Same Material," *Correlation*, 4, 1, May 1984, pp. 8-24.)

APPENDIX 3

Statistical Analysis

To observe planetary heredity, similarities of planetary positions have to be more frequent between the birth of parents and their children than expected. But exactly what does that mean, statistically speaking? First, we have to remember my previous observations on famous people. I described two categories of zones of the sky where the effect of the planet is different:

1) "Plus Zones," situated after the rise and the upper culmination;
2) "Minus Zones," situated in the remaining areas of the diurnal movement.

EXACT SIZE AND LOCATION OF RISE-CULMINATION ZONES
Before undertaking any statistical analysis of heredity experiments, we must carefully establish the exact size and location of the significant rise-culmination zones; we must then keep these zones constant **without any change** over all analyses. This recommendation is not at all superfluous. It is a common error in statistics to modify, more or less slightly, the variables under investigation in order to improve the significance of the results; this turns a perfectly nonsignificant observation into a very significant artifact.

Fortunately, my previous results with successful professionals allow me to establish these zones with clear objectivity. Look at the top graph of Figure 8, which was first published in 1960 in my work *Les Hommes & les Astres* (*Men and Stars*). It summarizes all the planetary observations I made on successful professionals, and gives the exact size and location of these zones. At that time I stated "both maxima occupy 1/9th of the diurnal movement and seem situated almost entirely after the horizon and the meridian but begin perhaps a little before." (*Les Hommes & les Astres*, p. 195).

More precisely, the rise zone corresponds to four 36-sectors, namely, 36, 1, 2, 3; and the culmination zone corresponds to four 36-sectors, namely, 9, 10, 11, 12. So from now on I call (+) the presence of a planet in these rise or culmination zones; and (−) the presence of a planet elsewhere in its diurnal movement. (This passage from *L'Hérédité Planétaire* , first published in 1966, is the first time I gave this precise astronomical description of what was later called by many authors the "Gauquelin plus zones"; see, for example, the glossary in every issue of the *Correlation* journal.) These definitions are used without change for my statistical analyses during **all** heredity experiments, including those after 1966, and **all** the celestial bodies of the solar system.

USE OF CHI-SQUARE TEST

The first stage of the work was to fill out index cards with all birth and planetary data for parents and child as illustrated in Figure 4.

For the second stage, that is, the statistical analysis of the data, the basic documents are arrays. These arrays allow for parent and child, all possible comparisons in 36 sectors for each celestial body studied. Special attention is given to the defined rise-culmination zones bounded by horizontal lines for the parent, vertical lines for the child, and cross squares for the encounter of both. Arrays allowed me to use a simple statistical test to determine the significance of the results. According to my hypothesis, all parent-child comparisons have to fall into one of the four categories indicated in Table 5.

TABLE 5
2X2 Statistical Test

		Children (+) Rise-culmination	**Children** (–) Elsewhere
Parents	(+) Rise-culmination	(+ +)	(+ –)
	(–) Elsewhere	(– +)	(– –)

The chi-square value indicates whether, overall, the hypothesis is supported or not:

$$\text{chi-square (1)} = \text{Sum of } \frac{(\text{obs} - \text{exp})^2}{\text{exp}}$$

(1) after chi-square means one degree of freedom. In this formula, obs is the observed frequency, and exp is the expected frequency, in each of the four categories. Table 6 shows the calculation of chi-square for the Moon for the whole heredity study.

TABLE 6
Example Calculation of Chi-square Value

		Children (+) zone	Children (–) zone	
Parents	(+) zone	849 782.21 +66.79 4,460.90 5.70	2,557 2,623.79 –66.79 4,460.90 1.70	3,406
	(–) zone	2,834 2,900.79 –66.79 4,460.90 1.54	9,797 9,730.21 +66.79 4,460.90 0.46	12,631
		3,683	12,354	16,037

Chi-square = 5.70 + 1.70 + 1.54 + 0.46 = *9.40.*
(significance level for one degree of freedom = .002)

In each box, from top to bottom, are the observed frequency, expected frequency, difference, square of difference, and chi-square value. The expected frequency is given by marginal total at right multiplied by marginal total at bottom divided by total at bottom right.

For example, the expected frequency for (+ +) is 3,406 X 3,683 / 16,037 = 782.21.

In this little 2 x 2 table, results in (+ +) and (– –) are in favor of the heredity hypothesis, whereas results in (+ –) and (– +) are against it.

After the chi-square value has been calculated, its significance (or probability) level is read from tables such as that given in part in Table 7. In my example, the value of chi-square(1) is 9.40, which is significant at the .002 level.

The individual values of chi-square(1) obtained for the moon, Venus, Mars, Jupiter and Saturn can be added to give chi-square(5), that is chi-square with five degrees of freedom. Table 7 gives also the probability level for several values of chi-square(5).

TABLE 7
Chi-square Values: Probability Levels

Probability	Interpretation	Chi-square(1)	Chi-square(5)
.25	not significant	1.32	6.63
.10	marginally sign.	2.71	9.24
.05	significant	3.84	11.07
.01	very significant	6.64	15.09
.001	"	10.83	20.52
.0001	"	15.14	25.75
.00001	"	19.51	30.86

A probability of .05 means that the result could have arisen by chance in 1 out of 20 experiments.

GRAPHIC PRESENTATION OF DATA

The chi-square test indicates only whether the planetary heredity hypothesis is confirmed or not. It does not describe how strong the planetary effect is or how it varies according to diurnal position. Therefore graphs are used to show this important information.

18-SECTOR DISTRIBUTION: EXPECTED FREQUENCIES

In the graphs showing diurnal distributions, the data used are not the observed frequencies, but the difference between observed and expected frequencies. All data which correspond to the graphs can be found in Appendix 4.

The expected frequencies were calculated as shown in the following example for sector 1 for the moon from Table 15A:

Total number of parent-child comparisons for the moon: 16,037.
Number of parents born with the moon in rise-culmination zones: 3,406.
Proportion between the two: 3,406 / 16,037 = 0.2124.
Number of children born with the moon in sector 1: 914.
Expected number of children born from parents with the moon in rise-culmination zones: 914 × 0.2124 = **194.**

The corresponding observed frequency is 208. The difference in observed-expected frequency is therefore 208 − 194 = +14.

PARENT-CHILD AND CHILD-PARENT DIRECTION

As previously explained, the data are analyzed in an array with two entries; the first entry gives results for the parent-child direction (left to right), which is the direction normally considered in heredity; the second entry gives results for the child-parent direction (top to bottom). In Figure 7 (and in no other figure) the planetary heredity results are given for both entries combined (Figure 7 corresponds to raw data published in Table 15C). This reduces chance fluctuations so that the general pattern becomes clearer. The individual results for each direction can be found in Appendix 4 (Tables 15A and 15B).

However, the intersection of (+) line and (+) column gives for the (+ +) zones the same number of planetary similarities in both directions. The addition of parent-child and child-parent directions remains thus in Figure 7 a graphic artifice; no probability can be calculated directly from it. Only the chi-square gives it correctly.

PARENTAL CATEGORIES: EXPECTED FREQUENCIES

The expected frequencies were calculated as shown in the following example from Table 11 for the category of (+ +) parents:

Number of cases with birth data for both parents: 3,487. The total for five planets combined is therefore 3,487 × 5 = 17,435.

Number of children born with a planet in rise-culmination zones, total for five planets combined: 3,976.

Proportion between the two: 3,976 / 17,435 = 0.2280.

Number of cases with both parents born with a planet in rise-culmination zones, total for five planets combined: 869.

Expected number with children born with a planet in rise-culmination zones, total for five planets combined: 869 × 0.2280 = **198.**

SIBLINGS

Results were analyzed in 2 × 2 tables similar to Table 5 with the younger sibling at the top and the elder sibling at the side. The question is: if one sibling is born with a planet in rise-culmination zones, does the other sibling show the same tendency? When there were more than two children in the family, I analyzed them in all possible combinations, following the chronological order of the siblings' births starting with the eldest. In every case I applied the same statistical treatment as I did when comparing parents and children. In this way, chi-square values were calculated for the two categories of parents: siblings from (+) parents and siblings from (−) parents (see Appendix 4, Table 12).

The expected frequencies for the chi-square test were calculated as shown in the following example for the moon and (+ +) siblings from (+) parents:

Number of comparisons between elder and younger siblings from (+) parents **and** from (−) parents: 6, 691.

Number of comparisons between elder and younger siblings from (+) parents only: 1,441.

Proportion between the two: 1,441 / 6,691 = 0.2154.
Number of (+) elder siblings from (+) **and** (−) parents: 1,545.

Number of (+) younger siblings from (+) **and** (−) parents: 1,562.

Expected number of (+) elder siblings from (+) parents: 1,545 × 0.2154 = 333.

Expected number of (+) younger siblings from (+) parents: 1,562 × 0.2154 = 336.

Consequently, at last, the expected moon frequency for (+ +) siblings, that is, (+) elder and (+) younger siblings, from (+) parents is 333 × 336 / 1,441 = 78.

The corresponding observed frequency is 92. The difference in observed-expected frequency is therefore 92 − 78 = +14.

UNDIFFERENTIATED SIBLINGS

I also analyzed planetary heredity between undifferentiated siblings, that is, without separating (+) and (−) parents. I found no significant results. For example, the observed moon frequency for (+ +) siblings from all parents combined is 353 (expected 360.68). The difference of −7.68 is not significant. The results are no better for Venus, Mars, Jupiter and Saturn. In other words, there is a specific link between siblings **only** if one takes care of the fact that they are born from (+) parents or from (−) parents, which is a rather logical result if we consider how the planetary heredity hypothesis was spelled out.

ONSET OF LABOR

The question is: if the parent (father or mother) was born with a planet in rise-culmination zones, is there a similar tendency at the onset of labor of the mother? Results were analyzed in 2 × 2 tables similar to Table 5 but with positions at the onset of labor taking the place of positions at the birth of the child. Naturally, I wanted to know if planetary positions at the onset of labor are linked to positions at the time of birth, in which case a significant planetary heredity at the onset of labor may have no meaning by itself. So I compared, in an array, the position of each planet at the onset of labor with its position at the time of birth. A rather clear relation was found between the two, simply because in many cases labor tends to take more or less the same time. But this did not significantly affect the rise-culmination zones, the observed and expected (+ +) cases being the same. Therefore, the previous results found at the onset of labor cannot be explained by an artifact due to the duration of labors in the study.

TIME OF DEATH

The question is: does a subject born with a planet in rise-culmination zones tend to die more frequently than chance under the same conditions? Results were analyzed using arrays comparing birth position with death position, using the same procedure as for onset of labor.

APPENDIX 4

Tables of Raw Results

To allow an independent check of my statistical calculations, this section gives the raw results used to produce each chi-square value and each graph, except those added since the first edition (for which see the relevant reference). It is divided into two categories of tables.

CATEGORY ONE:
RAW RESULTS USED FOR CHI-SQUARE TESTS
Tables give data which allow one to compute chi-square value according to the analysis I describe in Appendix 3 (pp. 74-80). This analysis is relevant for all the situations of planetary heredity I have studied in the present book. For each case, reference to a page of the book is given.

CATEGORY TWO: RAW RESULTS PLOTTED IN GRAPHS
(According to the astronomical conditions of the planetary heredity hypothesis, see Figure 5, p. 17.) Tables give observed and expected frequencies for every figure published with reference number, title and page of the book where it appears.

See Appendix 3 (pp. 74-80) for the calculation of expected frequencies.

Category One: Raw Results Used for Chi-Square Tests

TABLE 8

The Distance Factor: Chi-square Test by Planets (see p. 19)

	(+ +) Category			TOTAL			Chi
Planets	Obs	Exp	Difference	(+) Parents	(+) Children	General	**Square (1)**
Moon	849	782	+67	3,406	3,683	16,037	9.40
Venus	924	854	+70	3,694	3,706	16,037	9.80
Mars	916	831	+85	3,644	3,657	16,037	14.59
Jupiter	824	786	+38	3,541	3,560	16,037	3.02
Saturn	815	768	+47	3,505	3,512	16,037	4.80
						$chi^2(5)$	41.61
Mercury	809	817	− 8	3,558	3,684	16,037	0.14
Uranus	759	743	+16	3,082	3,868	16,037	0.54
Neptune	1,067	1,034	+33	4,288	3,867	16,037	1.90
Pluto	1,182	1,183	− 1	4,360	4,351	16,037	0.00
Sun	818	811	+ 7	3,438	3,784	16,037	0.09
						$chi^2(5)$	2.67

TABLE 9
Chi-square Tests as Function of the Epoch of the Child's Birth
(see p. 27)

Planets	(+ +) Category			TOTAL			Chi
	Obs	Exp	Difference	(+) Parents	(+) Children	General	Square (1)
	Births before 1931 - Paris						
Moon	347	293	+54	1,265	1,368	5,903	16.38
Venus	334	308	+26	1,340	1,355	5,903	3.82
Mars	323	295	+28	1,317	1,322	5,903	4.43
Jupiter	283	268	+15	1,315	1,202	5,903	1.40
Saturn	220	220	0	1,201	1,081	5,903	0.00
						$chi^2(5)$	26.03
	Births before 1938 - Parisian Suburbs						
Moon	274	255	+19	1,113	1,226	5,354	2.36
Venus	299	265	+34	1,163	1,222	5,354	7.02
Mars	325	293	+32	1,280	1,225	5,354	6.01
Jupiter	273	256	+17	1,145	1,196	5,354	1.90
Saturn	264	248	+16	1,168	1,138	5,354	1.64
						$chi^2(5)$	18.93
	Births after 1938 - Parisian Suburbs						
Moon	228	234	− 6	1,028	1,089	4,780	0.25
Venus	291	281	+10	1,191	1,129	4,780	0.57
Mars	268	243	+25	1,047	1,110	4,780	4.24
Jupiter	268	263	+ 5	1,081	1,162	4,780	0.17
Saturn	331	307	+24	1,136	1,293	4,780	3.29
						$chi^2(5)$	8.52

TABLE 10

Chi-square Tests according to the Sex of the Parents & Children (see p. 30)

Planets	(+ +) Category Obs	Exp	Difference	TOTAL (+) Fathers	(+) Children	General	Chi Square (1)
Moon	355	348	+7	1,557	1,632	7,308	0.24
Venus	411	376	+35	1,630	1,684	7,308	5.59
Mars	415	389	+26	1,712	1,660	7,308	2.97
Jupiter	369	346	+23	1,564	1,619	7,308	2.39
Saturn	368	338	+30	1,545	1,601	7,308	4.18
						chi²(5)	15.37
	Obs	Exp	Difference	(+) Mothers	(+) Children	General	Square (1)
Moon	494	435	+59	1,849	2,054	8,729	13.23
Venus	513	478	+35	2,064	2,022	8,729	4.35
Mars	501	442	+59	1,932	1,997	8,729	13.12
Jupiter	455	440	+15	1,977	1,941	8,729	0.90
Saturn	447	429	+18	1,960	1,911	8,729	1.24
						chi²(5)	32.84
	Obs	Exp	Difference	(+) Parents	(+) Sons	General	Square (1)
Moon	428	397	+31	1,739	1,889	8,270	3.93
Venus	503	463	+40	1,941	1,972	8,270	5.98
Mars	462	425	+37	1,873	1,875	8,270	5.49
Jupiter	449	413	+36	1,848	1,850	8,270	5.08
Saturn	424	398	+26	1,809	1,820	8,270	2.76
						chi²(5)	23.24
	Obs	Exp	Difference	(+) Parents	(+) Daughters	General	Square (1)
Moon	421	386	+35	1,667	1,797	7,767	5.35
Venus	421	391	+30	1,753	1,734	7,767	3.73
Mars	454	406	+48	1,771	1,782	7,767	9.41
Jupiter	375	373	+2	1,693	1,710	7,767	0.02
Saturn	391	369	+22	1,696	1,692	7,767	2.00
						chi²(5)	20.51

TABLE 11

Results according to Parental Categories (see p. 33)

Five planets added; for expected frequencies computation, see p. 34.

Category of Parents	Number of Comparisons	Planet in (+) Zones at the Child's Birth		
		Observed Frequencies	Expected Frequencies	Percentages
(+ +)	869	221	198	+ 12 %
(+ ?)	9,938	2,379	2,230	+ 7 %
(+ −)	6,114	1,507	1,394	+ 7 %
(− ?)	35,377	7,787	7,936	− 2 %
(− −)	10,452	2,248	2,384	− 6 %

TABLE 12

Siblings Chi-square Results (see p. 36)

Planets	(+ +) Category			TOTAL			Chi-square (1)
	Obs	Exp	Difference	Elder (+) Siblings	Younger (+) Siblings	General	
	Family (+)						
Moon	92	78	+ 14	333	336	1,441	4.62
Venus	103	94	+ 9	399	410	1,747	1.55
Mars	134	99	+ 35	427	416	1,784	20.55
Jupiter	122	97	+ 25	423	426	1,857	10.90
Saturn	115	84	+ 31	395	383	1,799	18.44
						$chi^2(5)$	56.06
	Family (−)						
Moon	261	283	− 22	1,212	1,226	5,250	2.90
Venus	283	265	+ 18	1,132	1,161	4,950	1.95
Mars	254	276	− 22	1,184	1,155	4,954	3.05
Jupiter	245	256	− 11	1,116	1,125	4,905	0.80
Saturn	207	229	− 22	1,076	1,042	4,896	3.43
						$chi^2(5)$	12.13

TABLE 13

Chi-square Result: Time of Onset of Labor (see p. 38)

Planets	(+ +) Category			TOTAL			Chi-square (1)
	Obs	Exp	Difference	(+) Parents	(+) Onset	General	
Moon	187	181	+ 6	813	845	3 794	
Venus	186	168	+ 18	950	669	3 794	
Mars	192	180	+ 12	841	813	3 794	
Jupiter	210	207	+ 3	848	927	3 794	
Saturn	233	231	+ 2	895	981	3 794	
Total	1,008	967	+ 41	4,347	4,235	18,970	2.90

TABLE 14

Chi-square Result: Time of Death of the Subject (see p. 39 & 40)

Planets	(+ +) Category			TOTAL			Chi-square (1)
	Obs	Exp	Difference	(+) Births	(+) Deaths	General	
Moon	246	237	+ 9	1,015	1,118	4,779	
Venus	285	264	+21	1,110	1,138	4,779	
Mars	233	228	+ 5	1,039	1,050	4,779	
Jupiter	225	227	− 2	1,036	1,045	4,779	
Saturn	219	209	+10	978	1,020	4,779	
Total	1,208	1,165	+43	5,178	5,371	23,895	2.74

Category Two: Raw Results Plotted in Graphs

The Five Significant Planets (see p. 21)

for Figure 7

(As especially mentioned in Appendix 3, p. 77, Table 15 is given in three parts. Table 15A: parent-child direction; Table 15B: child-parent direction; Table 15C: addition of both directions.)

TABLE 15A

Parent-child Direction

Sect	Moon		Venus		Mars		Jupiter		Saturn	
	Obs	Exp	Obs	Exp	Obs	Exp	Obs	Exp	Obs	Exp
1	208	194	275	239	236	199	198	185	193	184
2	241	211	249	233	237	222	202	202	220	206
3	195	188	230	230	219	234	178	197	168	173
4	169	180	227	234	219	216	182	193	174	181
5	174	177	198	210	216	208	193	186	210	191
6	209	190	224	196	218	208	209	205	204	189
7	192	190	181	176	229	217	200	198	182	182
8	175	184	166	184	182	208	197	194	192	176
9	178	184	176	189	205	211	180	178	202	200
10	198	193	159	174	196	197	219	197	177	189
11	195	196	188	184	191	185	211	218	196	207
12	172	184	185	203	158	176	183	200	192	203
13	175	189	211	198	179	186	194	200	212	205
14	177	182	194	190	157	173	170	200	191	210
15	179	194	199	204	187	184	197	194	206	199
16	192	187	202	209	198	193	216	192	199	213
17	193	189	204	212	225	227	206	204	208	197
18	184	194	226	229	192	200	206	198	179	200
N	3,406	3,406	3,694	3,694	3,644	3,644	3,541	3,541	3,505	3,505

TABLE 15B
Child-parent Direction

	Moon		Venus		Mars		Jupiter		Saturn	
Sect	Obs	Exp	Obs	Exp	Obs	Exp	Obs	Exp	Obs	Exp
1	213	192	275	244	245	214	201	201	211	200
2	210	199	229	219	273	232	221	203	205	189
3	225	216	232	225	201	196	208	213	175	180
4	196	206	201	223	177	207	187	187	172	175
5	210	205	199	189	180	197	201	202	221	192
6	206	197	214	193	200	193	215	196	204	200
7	227	212	199	185	200	199	194	197	161	185
8	200	207	191	185	177	193	189	187	187	194
9	201	206	183	175	181	193	227	208	179	182
10	195	204	159	182	195	186	197	195	207	205
11	202	213	176	181	181	185	207	216	189	206
12	224	209	179	186	204	216	168	182	208	192
13	197	193	181	196	208	210	189	189	191	186
14	191	205	214	206	204	208	206	207	202	197
15	191	202	207	217	220	205	208	196	217	211
16	190	194	220	222	209	199	181	196	202	212
17	202	211	212	243	194	213	180	187	192	201
18	203	212	235	235	208	211	181	198	189	205
N	3,683	3,683	3,706	3,706	3,657	3,657	3,560	3,560	3,512	3,512

TABLE 15C
Addition of Both Directions

	Moon		Venus		Mars		Jupiter		Saturn	
Sect	Obs	Exp	Obs	Exp	Obs	Exp	Obs	Exp	Obs	Exp
1	421	386	550	483	481	414	399	386	404	384
2	451	410	478	452	510	454	423	405	425	394
3	420	404	462	455	420	430	386	410	343	353
4	365	386	428	457	396	423	369	381	346	356
5	384	382	397	400	396	405	394	389	431	383
6	415	388	438	389	418	400	424	402	408	390
7	419	402	380	361	429	416	394	394	343	367
8	375	391	357	368	359	401	386	381	378	369
9	379	390	359	365	386	404	407	386	381	382
10	393	397	318	356	391	383	416	392	384	393
11	397	409	364	365	372	370	418	431	385	413
12	396	393	364	388	362	391	351	382	400	395
13	372	382	392	394	387	396	383	389	403	391
14	368	386	408	396	361	381	376	406	393	407
15	370	396	406	421	407	389	405	390	423	410
16	382	381	422	431	407	392	397	388	401	425
17	395	400	416	455	419	441	386	391	400	398
18	387	406	461	464	400	411	387	398	368	406
N	7,089	7,089	7,400	7,400	7,301	7,301	7,101	7,101	7,016	7,016

TABLE 16
Sex and Planetary Heredity
for Figures 11 & 12 (see p. 31-32)

Sect	Father		Mother		Son		Daughter	
	Obs	Exp	Obs	Exp	Obs	Exp	Obs	Exp
1	488	453	622	547	568	517	542	483
2	520	484	629	589	620	575	529	498
3	433	450	557	570	513	536	477	483
4	453	449	518	553	525	528	446	475
5	450	432	541	539	513	495	478	477
6	470	441	594	548	559	513	505	476
7	413	431	571	532	500	496	484	467
8	414	422	498	525	494	509	418	438
9	430	429	511	533	508	491	433	471
10	431	443	518	508	510	483	439	467
11	438	445	543	546	503	522	478	469
12	389	432	501	535	438	501	452	465
13	440	452	531	527	471	487	500	492
14	398	427	491	529	459	489	430	467
15	455	443	513	533	510	505	458	471
16	458	446	549	548	512	510	495	485
17	464	463	572	566	512	521	524	507
18	464	466	523	554	495	532	492	489
N	8,008	8,008	9,782	9,782	9,210	9,210	8,580	8,580

TABLE 17
Planetary Effect Vanishes with Surgical Intervention
for Figure 9 (see p. 26)

A
Hours of Birth

Hour	Freq.
0	24
3	20
6	26
9	38
12	38
15	17
18	22
21	30
N	215

B
Planetary Effect

Sect.	Obs	Exp
1	19	22
2	26	26
3	21	19
4	21	21
5	19	19
6	16	19
7	19	17
8	19	16
9	22	18
10	17	17
11	16	14
12	15	17
13	21	21
14	17	15
15	19	19
16	14	15
17	15	17
18	13	17
N	329	329

TABLE 18

Diurnal Rhythm & Planetary Effect Shift Ahead after 1938

for Figure 10 (see p. 28)

A

Hours of Birth

Hour	Before 1938	After 1938
0	1,108	451
3	1,262	500
6	1,280	498
9	1,305	505
12	1,185	432
15	997	386
18	992	416
21	1,041	434
N	9,170	3,622

B

Planetary Effect

	Before 1938		After 1938	
Sect.	Obs	Exp	Obs	Exp
1	763	698	347	303
2	772	704	377	369
3	672	684	318	338
4	657	690	314	314
5	664	659	327	313
6	721	647	343	341
7	679	641	305	321
8	604	642	307	303
9	600	627	342	336
10	668	663	283	289
11	722	724	259	266
12	640	688	250	277
13	686	693	283	283
14	619	679	270	276
15	661	686	307	289
16	717	709	290	286
17	735	734	301	296
18	727	739	260	283
N	12,307	12,307	5,483	5,483

TABLE 19
Parents with the Same Planetary Heredity: The Planetary Effect Increases for Children
for Figure 13 (see p. 35)

Sect. N°	(+ +)		(+ ?) & (+ −)		(− −)		(− ?)	
	Obs	Exp	Obs	Exp	Obs	Exp	Obs	Exp
1	55	48	1000	905	521	575	1971	2018
2	60	55	1029	964	636	656	2017	2067
3	41	50	908	921	627	605	2010	2010
4	56	50	859	905	606	601	1995	1961
5	50	48	891	876	571	578	1901	1911
6	58	49	948	889	569	595	1888	1929
7	58	46	868	870	546	557	1939	1938
8	43	45	826	856	548	544	1943	1915
9	50	47	841	867	562	570	1932	1901
10	47	46	855	857	545	558	1911	1897
11	49	49	883	892	592	586	1966	1963
12	32	48	826	871	607	573	1935	1909
13	43	46	885	886	560	556	1997	1997
14	36	45	817	864	562	545	1984	1944
15	46	47	876	882	580	565	1956	1965
16	53	49	901	897	590	584	1965	1979
17	47	50	942	930	594	596	2058	2065
18	45	51	897	920	636	608	2009	2008
N	869	869	16,052	16,052	10,452	10,452	35,377	35,377

TABLE 20
Planetary Heredity & Onset of Labor
for Figure 14 (see p. 38)

Sect.	1	2	3	4	5	6	7	8	9	10	11	12	N.
Exp	391	354	350	368	374	410	339	352	368	364	364	313	4,347
Obs	413	330	325	381	369	383	348	369	377	366	348	338	4,347

TABLE 21
Planetary Effect between Birth & Death
for Figure 15 (see p. 40)

Sect.	1	2	3	4	5	6	7	8	9	10	11	12	N.
Exp	448	431	432	431	438	412	413	418	422	450	447	436	5,178
Obs	480	448	437	449	434	414	399	413	309	423	455	427	5,178

TABLE 22

Relationship between Geomagnetic Activity and Planetary Effect
(Chi-square test)

for Figure 16 (see p. 49)

Ci	Categories Parent-child			
	(+ +)		(+ −)	
	Obs	Exp	Obs	Exp
0.0-0.6	2,309	2,376	7,476	7,439
0.7-1.3	1,485	1,438	4,542	4,501
1.4-2.0	385	365	1,064	1,142
N	4,179	4,179	13,082	13,082
Chi-square (2)	4.58, p = 0.1		5.89, p = 0.05	

TABLE 23

Planetary Heredity Increases for High Values of Ci

for Figure 17 (see p. 50)

Sect. N°	Children Born			
	Ci ≥ 1.0		Ci < 1.0	
	Obs.	Exp.	Obs.	Exp.
1	279	242	791	729
2	295	260	815	782
3	241	247	720	744
4	237	243	706	731
5	265	235	689	708
6	272	239	759	720
7	228	233	720	701
8	211	229	679	689
9	224	233	691	701
10	244	230	677	692
11	228	240	727	721
12	187	234	684	704
13	226	237	719	712
14	232	231	634	695
15	206	236	738	710
16	234	241	737	724
17	249	249	760	749
18	248	247	710	744
N	4,306	4,306	12,956	12,956

Bibliography

Annuaire du Bureau des Longitudes (published every year), Gauthier-Villars, Paris.

Bernard, Claude, *An Introduction to the Study of Experimental Medicine*, Dover, New York, 1957.

Bigg, E. K., "Lunar and Planetary Influences on Geomagnetic Disturbances," *Journ. Geophys. Res., 68*, 1963.

Brown, F.A., "How Animals Respond to Magnetism," *Discovery*, November 1963.

Chapman, S., *Magnetism and the Cosmos*, Oliver & Boyd, London, 1967.

Charles, E., "The Hour of Birth," *British Journal of Preventive & Social Medicine*, 7,43, 1953.

Choisnard, Paul, *La Loi d'Hérédité astrale*, Chacornac, Paris, 1919.

Cyran, W., *Geburtshilfe & Frauenheilkunde*, 10,667, 1950.

Dessagne, F. *Thèse de Médecine*, Paris, 1948.

Gauquelin, Michel, *L'Influence des Astres, Étude Critique et Expérimentale*, Dauphin, Paris, 1955.

___________, *Les Hommes & les Astres* (*Men and Stars*), Denoel, Paris, 1960.

___________, Die Planetare Hereditàt, *Zeischrift für Parapsychologie und Grenzgebiete der Psychologie*, 5, 2/3:168-193, 1961.

___________, "La latitude zodiacale," *Les Cahiers Astrologiques*, 92, May 1961.

___________, "L'hérédité astrale," *Les Cahiers Astrologiques*, 98, May/June 1962.

___________, *L'Hérédité Planétaire* , Denoel, Paris, 1966.

___________, "Note sur le rythme journalier du début des douleurs de l'accouchement," *Gyn. Obst.*, 66, Paris, 1967.

___________, "L'effet planétaire d'hérédité et le magnétisme terrestre," *Les Cahiers Astrologiques*, 134, 459-473, 1968.

___________, "L'effet planétaire d'hérédité en fonction de la distance de Vénus et de Mars a la Terre," *Les Cahiers Astrologiques*, 136, 581-569, 1968.

___________, "L'effet planétaire d'hérédité dans les secteurs du coucher et de la culmination inferieure," *Les Cahiers Astrologiques*, 134, 451-458, 1968.

___________, "Nombre relatif journalier des taches solaires et l'effet planétaire d'hérédité," *Les Cahiers Astrologiques*, 143, 259-262, 1969.

___________, "Un cas particulier dans l'effet planétaire d'hérédité, *Les Cahiers Astrologiques*, 141, 136-138, 1969.

___________, "Planetary Heredity, a Reappraisal on 50,000 Subjects," *New Birthdata Series*, Volume 2, Laboratoire d'etude des Relations entre Rythmes Cosmiques et Psychophysiologiques (LERRCP), Paris, 1984.

___________, "Profession & Heredity Experiments: Computer Reanalysis and New Investigations on the Same Material," *Correlation*, 4, 1, 8-24, May 1984.

___________, *The Truth About Astrology*, Blackwell, Oxford, 1983 (published in the United States as *Birthtimes*, Hill & Wang, New York, 1984).

___________, "On the lack of Positive Replication of Planetary Heredity," *Correlation*, 5,1, 36-42, 1985.

___________, *Cosmic Influences on Human Behavior*, Third Edition, Aurora Press, New York, 1985.

___________, "Bibliographic Chronology of Michel Gauquelin's Planetary Heredity Research," *Correlation*, 6, 1, 16-17, 1986.

___________, *Written in the Stars — The Best of Michel Gauquelin*, Aquarian Press, Wellingborough, U.K., 1988.

Gauquelin, Marie Françoise, "L'Heure de la Naissance," *Population*, 4,683, 1959.

Gauquelin, M. F. and Michel, *Methodes pour Etudier la Repartition des Astres dans le Mouvement Diurne*, Paris, 1957.

Gauquelin, Michel and Françoise, *Heredity Experiment, Series B*, 6 Volumes, LERRCP, Paris, 1970-1971.

___________, *Replication of the Planetary Effect in Heredity, Series D*, Volume 2, LERRCP, Paris, 1977.

Goehlert, V., *Biolog. Zentr. Blatt, 7*, 725, 1888.

Hosemann, H., *Zeitschrift für Geburtshilfe & Gynäkologie , 133*, 3,263, 1950.

Kepler, Johannes, *Harmonices Mundi*, 1619.

Kirschhoff, H., *Zentr. Blatt f. Gynäkologie, 3*, 135, 1935.

Knappich, Wilhelm, *Les Cahiers Astrologiques*, May 1960.

Krafft, Karl E., *Traité d'Astrobiologie*, Paris, Legrand, 1939.

Malek, J., *Ann. of the N.Y. Acad. of Sciences, 98*, 1042, 1962.

Merger, R., *Précis d'Obstretrique*, Masson, Paris, 1957.

Meyer, J., *Thèse de Médecine*, Lyon, 1956.

Ratcliff, J. D., *La Naissance*, Stock, Paris, 1953.

Reiter, R., *Z. Angew. Met., 1*, 289, 1953.

Rocard, Yves, *Le Signal du Sourcier*, Dunod, Paris, 1962.

Vaissiere, R., *Thèse de Médecine*, Toulouse, 1936.

Vignes, H., *Les Douleurs de l'Accouchement*, Masson, Paris, 1951.

Waldmeir, M., *The Sunspot Activity*, Zürich, 1961.

Index

G

H

I

J

K

L

M

N

O

P

Notes

Notes

Notes

Notes

GAUQUELIN PUBLICATIONS

LABORATOIRE D'ETUDE DES RELATIONS ENTRE RYTHMES COSMIQUES ET PSYCHOPHYSIOLOGIQUES

available from:
Michel Gauquelin, director
8, rue Amyot, 75005 Paris—France

SERIES A PROFESSIONAL NOTABILITIES

Volume 1: 2088 Sports Champions
Volume 2: 3647 Men of Science
Volume 3: 3439 Military Men
Volume 4: 2722 Painters & Musicians
Volume 5: 2412 Actors & Politicians
Volume 6: 2027 Writers & Journalists

SERIES B HEREDITY EXPERIMENT

Volume 1: Births from no. 1 to 5011
Volume 2: Births from no. 5012 to 9838
Volume 3: Births from no. 9847 to 13740
Volume 4: Births from no. 13741 to 17499
Volume 5: Births from no. 17500 to 21243
Volume 6: Births from no. 21244 to 24949

SERIES C PSYCHOLOGICAL MONOGRAPHS

Volume 1: Profession-Heredity, results of Series A & B
Volume 2: The Mars Temperament and Sports Champions
Volume 3: The Saturn Temperament and Men of Science
Volume 4: The Jupiter Temperament and Actors
Volume 5: The Moon Temperament and Writers

SERIES D SCIENTIFIC DOCUMENTS

Volume 1: The Planetary Factors in Personality
Volume 2: New Replication of the Planetary Effects in Heredity
Volume 3: Statistical Tests of Zodical Influences: Profession-Heredity
Volume 4: The Venus Temperament, a Tentative Description
Volume 5: Sun, Mercury, Uranus, Neptune, Pluto: Profession & Heredity
Volume 6: The Mars Effect & Sports Champions: A New Replication
Volume 7: Traditional Symbolism in Astrology & Character Traits
Volume 8: Zodiac & Character Traits
Volume 9: Murderers & Psychotics (with 7,035 Birth Data)
Volume 10: Report on American Data (with 1401 Birth Data and 5456 Character Traits of Successful Americans)

NEW BIRTHDATA SERIES (1984)

Volume 1: 2145 Physicians, Army Leaders, Top Executives
Volume 2: Planetary Heredity: A Reappraisal on 50,000 Subjects with Publication of 15,000 Birth Data
Volume 3: 1540 Authors, Artists, Actors, Politicians, Journalists

---

Other ACS Books by Michel Gauquelin

The Cosmic Clocks
The Gauquelin Book of American Charts

Also by ACS Publications, Inc.

All About Astrology Series
The American Atlas: US Latitudes and Longitudes, Time Changes and Time Zones (Shanks)
The American Book of Nutrition & Medical Astrology (Nauman)
The American Book of Tables
The American Ephemeris Series 1901-2000
The American Ephemeris for the 20th Century [Midnight] 1900 to 2000
The American Ephemeris for the 20th Century [Noon] 1900 to 2000
The American Ephemeris for the 21st Century 2001-2050
The American Heliocentric Ephemeris 1901-2000
The American Midpoint Ephemeris 1986-1990 (Michelsen)
The American Sidereal Ephemeris 1976-2000
A New Awareness (Nast)
Asteroid Goddesses (George & Bloch)
Astro-Alchemy: Making the Most of Your Transits (Negus)
Astrological Games People Play (Ashman)
Astrological Insights into Personality (Lundsted)
Astrology: Old Theme, New Thoughts (March & McEvers)
Basic Astrology: A Guide for Teachers & Students (Negus)
Basic Astrology: A Workbook for Students (Negus)
Beyond the Veil (Laddon)
The Body Says Yes (Kapel)
Comet Halley Ephemeris 1901-1996 (Michelsen)
Complete Horoscope Interpretation: Putting Together Your Planetary Profile (Pottenger)
Cosmic Combinations: A Book of Astrological Exercises (Negus)
Easy Tarot Guide (Masino)
Expanding Astrology's Universe (Dobyns)
The Fortunes of Astrology: A New Complete Treatment of the Arabic Parts (Granite)
The Gold Mine in Your Files (King)
Hands That Heal (Burns)
Healing with the Horoscope: A Guide to Counseling (Pottenger)
The Horary Reference Book (Ungar & Huber)
Horoscopes of the Western Hemisphere (Penfield)
Houses of the Horoscope (Herbst)
Instant Astrology (Orser & Brightfields)
The International Atlas: World Latitudes, Longitudes and Time Changes (Shanks)
Interpreting Solar Returns (Eshelman)
Interpreting the Eclipses (Jansky)
The Koch Book of Tables
The Mystery of Personal Identity (Mayer)
The Only Way to...Learn Astrology, Vol. I Basic Principles (March & McEvers)
The Only Way to...Learn Astrology, Vol. II Math & Interpretation Techniques (March & McEvers)
The Only Way to...Learn Astrology, Vol. III Horoscope Analysis (March & McEvers)
Past Lives Future Growth (Marcotte & Druffel)
Planets in Combination (Burmyn)
The Psychic and the Detective (Druffel with Marcotte)
Psychology of the Planets (F. Gauquelin)
Secrets of the Palm (Hansen)
Seven Paths to Understanding (Dobyns & Wrobel)
Spirit Guides: We Are Not Alone (Belhayes)
Stalking the Wild Orgasm (Kilham)
Twelve Wings of the Eagle (Simms)
Tomorrow Knocks (Brunton)
12 Times 12 (McEvers)